From Colonization to Domestication

From Colonization to Domestication

Population, Environment, and the Origins of Agriculture in Eastern North America

D. SHANE MILLER

THE UNIVERSITY OF UTAH PRESS

Salt Lake City

The Defiance House Man colophon is a registered trademark
of The University of Utah Press. It is based on a four-foot-tall
Ancient Puebloan pictograph (late PIII) near Glen Canyon, Utah.

LIBRARY OF CONGRESS CATALOGING-IN-PUBLICATION DATA

Names: Miller, D. Shane (Darcy Shane), 1982– author.
Title: From colonization to domestication : population, environment, and the
    origins of agriculture in eastern North America / D. Shane Miller.
Description: Salt Lake City : The University of Utah Press, [2018] | Includes
    bibliographical references and index. |
Identifiers: LCCN 2017055320 (print) | LCCN 2017057345 (ebook) |
    ISBN 9781607816171 ( ) | ISBN 9781607816164 (cloth)
Subjects: LCSH: Paleo-Indians—Agriculture—Southern States. | Indians of
    North America—Agriculture—Southern States. | Agriculture,
    Prehistoric—Southern States. | Agriculture—Southern States—Origin. |
    Indians of North America—Southern States—Antiquities. | Excavations
    (Archaeology)—Southern States. | Environmental archaeology—Southern
    States. | Social archaeology—Southern States. | Southern States—Antiquities.
Classification: LCC E78.S65 (ebook) | LCC E78.S65 M555 2018 (print) |
    DDC 975.004/97—dc23
LC record available at https://lccn.loc.gov/2017055320

Printed and bound in the United States of America.

To my Mom, for all of those trips to the public library

Contents

Figures

Tables

Acknowledgments

If you know anything about me, it should be apparent that I'm going to start by thanking my mom. We've been through a lot, and she's always been in my corner. She's my hero. I'm also pretty lucky in that I have a very supportive family, including my stepdad, Jim, and little brother, Jesse. In my mind, they're taking bets on whether I'm going to send a copy of this book to the Greeneville, Tennessee, Public Library.

As soon as I got to Tucson to start the doctoral program at the University of Arizona, I knew I was in good hands. So much so that when it came time to pick an advisor I had multiple options. Vance Holliday and Steve Kuhn agreed to cochair my committee, and they have been fantastic mentors. Whether the subject was stone tools, soils, theory, music, or life, they have been everything I could have asked for and more. I can walk into their offices and talk to them about anything. I am proud to be one of their students, and I'm glad they pushed me to write a dissertation that could be turned into a book manuscript.

At one time I was really struggling with finding a dissertation topic and questioning if what I was doing was worthwhile. Then I took my comprehensive exams, and Mary Stiner asked me a question that launched me on a much more fulfilling path. She pushed me to be creative, and I can't thank her enough for that.

David Anderson has been one constant throughout my career. From my very first day as a master's student at Tennessee to the completion of my dissertation at Arizona, he's been there for me. He made Southeastern Archaeology interesting, fun, and something that I wanted to pursue with reckless abandon.

It is not possible to survive as an academic without the help of the departmental administrative staff. Catherine Lehman, Norma Maynard,

Ben Benshaw, Ann Samuelson, Veronica Peralta, Debbie Vickers, and Kathy Elliot are the bureaucratic ninjas who can find a way to cut through even the thickest of red tape.

The research in this book was supported by a dissertation improvement grant from the SRI Foundation, with help from the McClung Museum of Natural History and Culture and the Tennessee Division of Archaeology. In particular, I'd like to thank Carla Van West, Jeff Chapman, Lynne Sullivan, Bobby Braly, Jessica Dalton Carriger, Mike Moore, Suzanne Hoyal, John Broster, Aaron Deter-Wolf, Ted Wells, and Scott Meeks for all of their help.

Two other individuals helped me get this research project off the ground. Tom Pertierra gave me much-needed advice on my budget and DuVal Lawrence spent several days with me measuring projectile points in the basement of the McClung Museum. I would call them the gold standard for the phrase "avocational archaeologists." I'm glad they call me their friend.

I firmly believe in the old adage "you are who you hang with." Sometimes you have control over this. Other times you just happen to be in the right place surrounded by the right people at the right time. The latter happened to me when I started graduate school at the University of Arizona in 2007, where I found myself in a classroom with Derek Anderson, Randy Haas, Adam Foster, Liye Xie, Meagan Trowbridge, Meredith Reifschneider, Lauren Hayes, Angela Storey, and a bunch of other folks. They challenged me intellectually and made it all fun. I consider these folks some of my closest, dearest friends.

As I was starting the doctoral program at Arizona, Thad Bissett and Stephen Carmody were beginning the doctoral program at Tennessee. Coincidentally, I was the graduate student who showed them around South Stadium Hall when they visited Knoxville during my last semester there. We stayed in touch, continued to talk about archaeology, eventually began doing projects together, and they were more than supportive when my interests shifted from the end of the Ice Age to the mid-Holocene. I hope to pester them with questions about Archaic hunter-gatherers for a very long time.

The people who have helped me over the years are legion, and it would take me at least five additional pages to list them all. This is largely a product of my grad school experiences at Tennessee and Arizona and working at a school like Mississippi State University. All three schools are populated with great people, and I consider myself very lucky to have made so many wonderful friends at these places.

I'd also like to thank Reba Rauch, Phil Carr, and Bruce Winterhalder for all of the great advice and constructive criticism on how to turn my dissertation research into a book manuscript.

It is a great honor, of which I hope to prove myself worthy, to have received the Don D. and Catherine S. Fowler Prize for this book. I am grateful to Don and Catherine Fowler for their support, and to the prize committee for selecting my manuscript.

Finally, I owe Kerri Mathews a lot of gratitude for being such a great partner and for being supportive when I go full nerd.

CHAPTER 1

Behavioral Ecology
and the Origins of Agriculture

As an undergraduate in Introduction to Archaeology, the peopling of the Americas fascinated me. The question was *big*, with many facets: Who were these people, how did they get to North America, and most controversially, *when* did they arrive? These questions captured my imagination because they were big in other ways. Two ice sheets covered most of what is now Canada, the Great Salt Lake would have been totally encapsulated by the much larger Lake Bonneville, the Floridian Peninsula would have been twice as wide due to much lower sea levels, and one could have witnessed icebergs floating off the coast of South Carolina (Holliday and Miller 2013).

As a kid who grew up in the southeastern United States, what struck me even more was imagining my surroundings covered by the same forests that now cover Quebec but with megafauna, including mastodons and sloths (Delcourt and Delcourt 1985; Meltzer 2009). The earliest Native Americans would have arrived into this alien world, and they perhaps irrevocably changed it by hunting many of these species to extinction, a debate that also grabbed my interest (Fiedel and Haynes 2004; Grayson and Meltzer 2003, 2004, 2015). Did humans take big-game hunting to a new level, and if so, could they have erased more than 30 species from existence despite being relatively few in number? It was the ultimate whodunit.

As a graduate student, one of my favorite courses was Southeastern Archaeology, which I took mostly to flesh out my knowledge of the region. I honestly did not expect to engage with much beyond the parts of the course that covered the Late Pleistocene and Early Holocene archaeological record. I overcame my paleo hubris a little more with each class, with the tipping point being the week we talked about the origins of agriculture—in particular, the uniqueness of eastern North America on the world stage. It is one of only a handful of places in the world where people unequivocally domesticated plants (Smith 1987, 1992, 2001).

While my interests over the course of my graduate career focused mostly on the colonization of the Americas, I continued to follow recent developments on the origins of agriculture in eastern North America, including articles by Bruce D. Smith, Kristen J. Gremillion, and Douglas J. Kennett and Bruce Winterhalder's (2006) *Behavioral Ecology and the Transition to Agriculture*. At the encouragement of my doctoral committee, I pursued my interests in the origins of agriculture as well as the peopling of the Americas. As a result, a rather straightforward yet far-reaching question emerged that allowed me to connect my two archaeological interests: How and why did Native Americans go from hunting mastodons to planting sunflowers in eastern North America?

The Origins of Agriculture

Beginning at least 14,000 years ago with the domestication of the gray wolf, modern humans have been responsible for artificially selecting traits in other species for their own benefit (Clutton-Brock 1999; Germonpre et al. 2012; Leonard et al. 2002). In at least six different locations worldwide, humans independently domesticated a variety of other plant and animal species, which sparked a radical shift in subsistence economies—from hunting and gathering to the near complete reliance on domesticated species (Barker 2006; Kennett and Winterhalder 2006:2; Price 2009; Figure 1.1). From these hearths of domestication, agriculture spread rapidly to encompass almost the entire planet. This transition irrevocably changed human dietary choices and generated long-term consequences for the physical environment and human nutrition that extend to the present day (Diamond 2002; Kennett and Winterhalder 2006; Redman 1999; Ungar and Teaford 2002; Zeder 2011). This study examines the possible roles of population pressure and resource imbalance in one of these hearths of domestication—eastern North America.

The shift from foraging to farming also coincided with major changes in technology and social organization (Barker 2006; Kennett and Winterhalder 2006:2) and led to irrevocable changes to physical environments, creating anthropogenic landscapes that are profoundly shaped by agri-

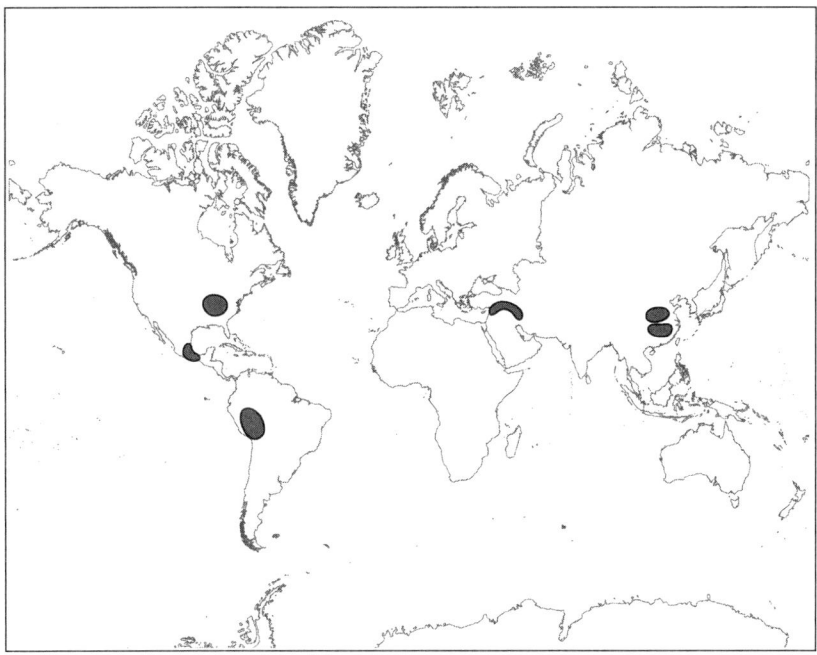

FIGURE 1.1. The locations of the earliest independent centers for plant and animal domestication. (Base Map: Bing Aerial Imagery provided by ESRI ArcMap 10)

cultural practices (Blondel 2006; Delcourt and Delcourt 2004; Mann 2006). In addition to changing the physical landscape to the point of degradation in some instances, the transition to agriculture is ultimately responsible for a less diverse diet (Milton 2002), increased evidence for skeletal pathologies related to nutritional deficiencies and increased workloads (Cohen and Armelagos 1984), higher probabilities for exposure to zoonotic and crowd-borne illnesses (Diamond 2002; Larsen 2002), and a variety of diseases that impact people today, including hypertension, diabetes, and heart disease (Eaton et al. 1988).

Given the negative consequences, why did people not simply continue to hunt and gather as a primary mode of subsistence? Jared Diamond (2002:701) famously posed the question somewhat differently, asking, "Why did food production eventually outcompete the hunter-gatherer lifestyle over almost the whole world, at the particular times and places that it did, but not at earlier times?" This places the process in historical context, demanding attention to conditions at particular places and times.

Why Become a Farmer?

Starting with Edward Tylor's (1871) classification of human societies into three levels: savagery, barbarism, and civilization, several generations of researchers have been interested in the transition from a hunting-and-gathering mode of subsistence to one that relies on agriculture. In the 1930s, V. Gordon Childe argued that a Neolithic revolution occurred at the end of the last Ice Age. In his oasis hypothesis, Childe (1936) contends that drier conditions drove animals and people to oases and river valleys, which would have prompted a codependent relationship. However, climate reconstructions have failed to support this hypothesis (Barker 2006:16–17).

Subsequently, Robert J. Braidwood (1960), after an extensive survey in the Zagros Mountains in Iraq, argued that the earliest evidence for agricultural societies did not occur near oases or major rivers but in the "hilly flanks" of the area known as the Fertile Crescent. Braidwood, however, approached the transition to agriculture from a culture-historical perspective and attributed the transition to agriculture to "cultural" factors (Barker 2006:26). In the ensuing years, the question of how and when human subsistence changed at this key juncture has been one of the most researched and debated issues in archaeological research.

Lewis R. Binford (1968) argued that increasing population pressure resulting from ameliorating climatic conditions at the end of the Pleistocene might be responsible for the appearance of domesticated plants and animals. Specifically, he cited an increase in sea level and population growth that likely increased population density in the Near East, which brought human groups closer to the carrying capacity of the regional environment. This in turn prompted out-migration to less populated regions and subsequent resource intensification. Kent V. Flannery (1969) applied a systems approach in an attempt to demonstrate that the transition from hunting and gathering to agriculture was the result of a series of positive feedback loops. Most notably, he contends that Late Pleistocene hunter-gatherers were already in the process of expanding their diet and utilizing a wide range of plants and animals.

Binford's (1968) and Flannery's (1969) hypotheses have been subsequently supported by regional zooarchaeological analyses (Stiner 2001) and by the discovery of early Epipaleolithic sites with diverse paleo-botanical assemblages, such as Ohalo II on the Sea of Galilee (Nadel and Werker 1999). With changes in climate at the end of the Pleistocene, hunter-gatherers responded by increasing their reliance and investment in harvesting, processing, and managing wild cereals. Binford (1968) and Flannery (1969) both argued that this type of investment would have been most necessary in marginal environments where hunter-gatherers were operating at or near the carrying capacity of their environment.

In a more contemporary formulation of the broad-spectrum revolution model of Binford (1968) and Flannery (1969), Michael Rosenberg (1998) hypothesized that as population density increases, groups will occupy spatially discrete and highly predictable resources for progressively longer periods of time. One by-product of this pattern is that in order to stay in such locations, groups must abandon less productive areas as well as locations that are too costly to defend. He argues that an additional outcome in many cases is that decreasing residential mobility fosters innovation to allow for survival in more spatially restricted and/or marginal areas. One such innovation may have been the inception of domesticated plants and animals.

On the other hand, David Rindos (1984) argues that the transition to agriculture in the Near East was likely the result of a coevolutionary relationship between people and the plants and animals they eventually domesticated. And Melinda Zeder (2011, 2012, 2015) argues that this provides an example of gradual niche construction (e.g., Odling-Smee et al. 2013). Bruce D. Smith (1987, 1992, 2001, 2011, 2015) has argued for a similar pathway to domestication in eastern North America and, like Zeder, has borrowed extensively from niche construction theory.

Barbara Stark (1986) and, later, Kennett and Winterhalder (2006) classified the various models for plant and animal domestication into three types. Binford's (1968) and Flannery's (1969) hypotheses would be categorized as a "push" model, where demographic or resource stresses are the catalysts for changes in subsistence. The second type is described as a "pull" model that relies on climate change and an increase in the availability of certain resources as a causal factor leading to a co-dependent relationship between plants and people. One such example is Childe's (1936) oasis hypothesis for the appearance of domesticated

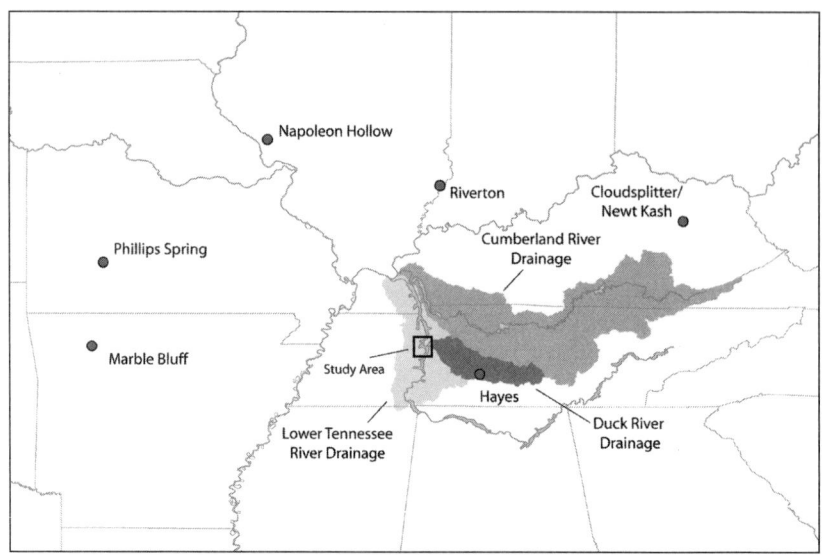

FIGURE 1.2. The locations of archaeological sites containing the oldest dated domesticated seeds in eastern North America with the Cumberland, lower Tennessee, and Duck River drainages highlighted. The primary location of this study (Benton and Humphreys counties) is also indicated.

TABLE 1.1. Earliest dated domesticated plant remains in eastern North America.[1]

Sites	Species	Age (^{14}C BP)	Age Range (cal BP)[2]	Lab Number
Phillips Spring, MO	Pepo squash (*C. pepo ssp. ovifera*)	4440±75	5342–4920	Beta-47293
Hayes, TN	Sunflower (*H. annuus*)	4265±60	5081–4640	Beta-45050
Napoleon Hollow, IL	Marsh-elder (*I. annua*)	3920±40	4562–4290	Beta-216463
Riverton, IL	Chenopod (*Ch. berlandieri*) ("naked")	3490±40	3919–3693	Beta-253114
Cloudsplitter, KY	Chenopod (*Ch. berlandieri*) (thin-testa)	3450±150	4195–3431	Beta-11348
Riverton, IL	Chenopod (*Ch. berlandieri*) (thin-testa)	3440±40	3882–3641	Beta-253117
Newt Kash, KY	Chenopod (*Ch. berlandieri*) (thin-testa)	3400±150	4134–3396	Beta-11347

[1] Adapted from Smith and Yarnell (2009:6564).
[2] Calibrated with OxCal 4.1 using the IntCal 09 curve (Bronk Ramsey 2009).

plants and animals in the Fertile Crescent as well as Zeder's (2011) recent niche construction hypothesis in the same region.

Finally, a third group of models focuses on socioeconomic or social factors. One example is Brian Hayden's (1992, 1995) hypothesis that plant and animal domestication may have occurred as a result of certain key individuals' desire to acquire resource surpluses in order to increase social status.

Plant Domestication in Eastern North America

Eastern North America is one of the areas of the world where indigenous plants were independently domesticated (Smith and Yarnell 2009; Kennett and Winterhalder 2006). The primary domesticates for this region include a variety of seed-bearing annuals, such as goosefoot (*Chenopodium berlandieri*) and sunflower (*Helianthus annuus*), that were domesticated ~5000–3800 cal BP. The best documented "earliest" dates occur at six sites: the Phillip Springs site in southwestern Missouri, the Hayes site in central Tennessee, the Napoleon Hollow and Riverton sites in southern Illinois, and Newt Kash and Cloudsplitter Rockshelters in Kentucky (Smith and Yarnell 2009:6561–6562; Figure 1.2; Table 1.1).

Beginning with the pioneering work of Volney Jones (1936), a considerable amount of research has been focused on documenting the timing (and, to a lesser degree, the context) of plant domestication in eastern North America. Subsequent interest in this topic has been pushed by a handful of individuals, one of the most notable being Bruce D. Smith (1987, 1992, 2007, 2011), who argued that plant domestication happened so much later in eastern North America compared to other *domestication hearths* because the environmental preconditions were not in place until the mid-Holocene. It was during this time that the major rivers across eastern North America stabilized and began aggrading, which created the meander belt topography that typifies eastern North America presently. The formation of more ecologically diverse floodplain habitats, accompanied by a decrease in precipitation, caused a deterioration of upland habitats and created a push-pull effect that prompted groups to settle more intensively in riverine environments. A coevolutionary relationship began to emerge between people and a set of plants that are adept at exploiting disturbed ground. More recently, Smith (2011:S482) has taken his arguments a step further, stating that

> Eastern North America, arguably the best-documented region case study available, does not provide much support for general models including those of human behavioral ecology that incorporate environmental downturn, external environmental stress, population growth, landscape packing, constructed resource zones, and carrying-capacity imbalance or resource scarcity in explaining the initial domestication process.

In addition to riverine habitats, there is also evidence for the use of domesticated plants in upland settings—most notably, on the Cumberland Plateau in central Kentucky (Smith and Cowan 1987; Gremillion 1996, 2002; Gremillion et al. 2008) and the Ozarks in Arkansas (Fritz 1990, 1997). Gremillion (2004) contended that the hunter-gatherers in the uplands during the Late Archaic and Early Woodland periods were low-level food producers (e.g., Smith 2001) and that the domesticates they were using represented a relatively low-return food item. However, based on pollen and seed remains from human coprolite samples, she argued that a temporal lag existed between the time food items were collected (late summer and early fall) and when they were consumed (late winter and early spring). In her view, this provided hard evidence that the stored seeds provided a food source in times of seasonal scarcity, when the time spent searching for alternative food sources was allocated to processing and consuming the seeds from domesticates. David W. Zeanah (2016) has also recently argued that the shift to cultivating domesticated plants in eastern North America is consistent with resource intensification as a result of population growth, which appears to coincide with an increase in the frequency of radiocarbon dates (e.g., Weitzel and Codding 2016).

Furthermore, Gremillion and colleagues (2008) found that the limestone benches in the escarpment of this region may have provided an optimal habitat for horticulture that could have been further augmented by controlled burning. Under these circumstances, the best subsistence strategy was to invest in weedy plants that could quickly exploit openings in the forest canopy and then store seeds in advance of anticipated seasonal resource shortfalls. Moore and Dekle (2010) suggest that the emergence of low-level food production in this area came about as a result of intensive seed, nut, and shellfish processing that began during the mid-Holocene. However, the upland populations had little to no access to riverine resources and consequently placed more emphasis on storing nuts and seeds.

Smith's view is in stark contrast with Gremillion's (2004) argument for agriculture's origins in eastern North America, which she largely derived from the expectations of the diet breadth model. Building on previous arguments by Winterhalder and Goland (1997), Gremillion contends that the seedy annuals in eastern North America were low-ranked resources. We should expect to see them used in two contexts: 1) if their profitability improved, perhaps as the result of technological innovations to make harvesting more efficient, or 2) where more highly ranked resources become scarce. Gremillion asserts that food storage and the inclusion of low-ranked seedy annuals in the diet was originally a strategy to combat seasonal food shortages. In this instance, Gremillion (2004) finds that, if the modern distribution of these species is a proxy for prehistoric ranges, most of the domesticated species utilized by the groups in the uplands of eastern Kentucky were well outside their native range.

Whereas Smith (2011) is skeptical that a whole array of variables could be causal explanations for why agriculture appeared when and where it did, Gremillion (1996, 2004; Gremillion and Piperno 2009) frequently makes use of models from human behavioral ecology (particularly the diet choice and central place foraging models) as an interpretive framework to understand the microeconomic decisions available to prehistoric hunter-gatherers and early farmers. More broadly, she also examines regional spatial trends, finding that domesticated plants in an archaeological context appear in the mid-continent—in particular, the lower Illinois River valley (e.g., Asch and Asch 1985), central Tennessee (Crites 1993), eastern Tennessee (Chapman and Shea 1981), and eastern Kentucky and the Ozarks (Cowan 1985; Gremillion 2004; Yarnell 1978). The spatial extent of these locations roughly corresponded to isotherms of the average number of days annually in which temperatures drop below 32° (Gremillion 2002:497; Figure 1.3). She argues that this further supports the hypothesis that there is some relationship between food

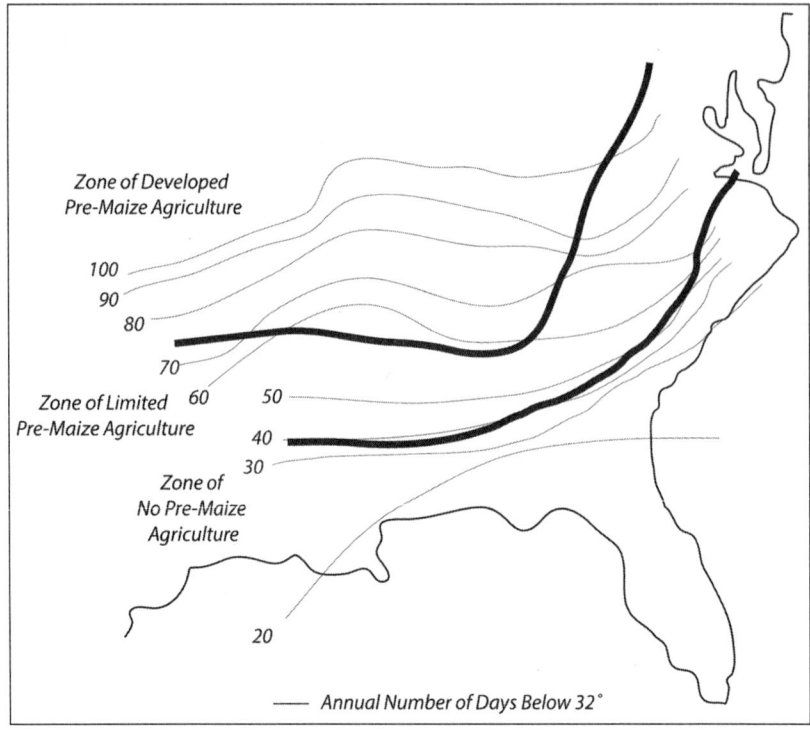

Zone of Developed
Pre-Maize Agriculture

100
90
80
70
Zone of Limited   60   50
Pre-Maize Agriculture
40
30
Zone of
No Pre-Maize
Agriculture

20

—— Annual Number of Days Below 32°

FIGURE 1.3. The distribution of pre-maize agriculture with isotherms indicating the annual number of days below 32° (adapted from Gremillion 2002:497).

production and seasonal shortfalls and that while Smith's model hinges on enriched river valleys, the distribution of sites and variation in their landscape contexts lead her to conclude that agriculture could have developed independently in other areas away from the major river valleys of the mid-continent (Gremillion 2002:492–493).

While Gremillion has focused on microeconomic and broader spatial trends, she has also called into question the degree to which subsistence has been stable over the Holocene (Gremillion 1996:99). The hypothesis that subsistence remained stable over the course of the Holocene gained traction first with Joseph R. Caldwell's (1958) primary forest efficiency model, and some versions of it are still popular (e.g., Anderson and Sassaman 2012:101–107). This view, at least in regard to subsistence, is still held by Smith (2011:482), who contends that in the Late Archaic period, "subsistence economies have remained stable over long periods of time."

However, Gremillion (1996:99) argues that this "transformation of inferences into assumptions may further reinforce our ignorance by discouraging the formulation of alternative hypotheses." She then cites Paul S. Gardner's (1997) hypothesis that mid-Holocene warming increased the abundance of oak and hickory as well as reduced the frequency of mast failure. This in turn prompted specialization, which is reflected in high proportions of oak relative to other species in the mid-continent during the mid-Holocene, a view that has been central to more recent studies of Middle Archaic subsistence (Bissett 2010; Carmody 2009, 2010; Moore and Dekle 2010). Consequently, Gremillion (1996: 111) finds that there is a need for the systematic gathering of archaeobotanical, faunal, and archaeological data that crosscut environmental boundaries at multiple scales of analysis (see also VanDerwarker and Peres 2010).

Based on limited evidence, there are several major trends that may illustrate the dynamics of subsistence change in the periods prior to plant domestication. While evidence for morphological changes in plants prior to ~5,000 cal BP has not yet been reported (Smith and Yarnell 2009), despite research programs geared toward capturing and analyzing organic remains (e.g., Asch et al. 1972; Marquart and Watson 2005), researchers have long observed evidence for significant and widespread changes in residential mobility, technological organization, and resource use during the Middle Holocene in the mid-continent (Amick and Carr 1996; Anderson et al. 2007; Brown and Vierra 1983; Sassaman 2010). This includes an increasing frequency of deer (*Odocoileus virginianus*) in archaeological faunal assemblages (Styles and Klippel 1996) and the appearance of freshwater shell middens in the major drainages of the mid-continent, including the Tennessee and Cumberland Rivers (Anderson et al. 2007; Bissett 2014; Claassen 1996; Meltzer and Smith 1986; Morse 1967; Sassaman 2010; Smith 1986; Steponaitis 1986).

Moreover, others contend that there is evidence for an increasingly diverse diet in the Late Paleoindian period based on the wide array of floral and faunal remains from sites such as Shawnee-Minisink in Pennsylvania (Dent 2007; Gingerich 2007, 2011) and Dust Cave in Alabama (Hollenbach 2007, 2009; Walker et al. 2001). Some have even questioned the assumption that big-game hunting specialists ever existed in eastern

North America, asserting that the earliest populations had a much more generalized diet (Grayson and Meltzer 2003, 2004, 2015), although this has generated much debate (Fiedel and Haynes 2004; Surovell and Waguespack 2008). Consequently, the questions of where, when, and why the trajectory leading to the domestication of plants and the adoption of agriculture began in eastern North America is unclear. To address this shortcoming requires pulling together disparate datasets that can provide information about prehistoric subsistence and a theoretical construct that allows us to make inferences about economic decisions made by prehistoric people.

Human Behavioral Ecology, *Homo Economicus*, and the Palimpsest Problem

While many models have been developed to explain why agriculture developed where and when it did, attempts to formally test them are few and far between, especially in eastern North America (Gremillion 2002). Kennett and Winterhalder (2006) argue that one shortcoming is the lack of an appropriate framework for interpreting changes in subsistence related to this transition. To this end, they advocate an approach derived from human behavioral ecology, which benefits from "its ability to carry into hypothesis generation a wide variety of postulated sources of causation—global climate change to the aggrandizement of dominant individuals" (Kennett and Winterhalder 2006:10).

Behavioral ecology grew out of research on evolutionary biology and animal behavior in the 1960s and 1970s. It was subsequently adopted by anthropologists and archaeologists, initially as a way to interpret subsistence decisions by hunter-gatherer groups but has since been applied to a much wider array of behaviors and contexts (Borgerhoff Mulder and Schacht 2012; Winterhalder and Smith 2000).

Broadly speaking, practitioners of behavioral ecology seek to understand how human behavior is shaped by economic and, ultimately, evolutionary influences (Smith et al. 2001:128). They commonly assume that closer-to-optimal alternatives, defined in terms of currencies such as energy or time, will tend to become more common over time at the expense of less efficient strategies (Kelly 2013). In other words, Borgerhoff Mulder and Schacht (2012:2) state, "Behavioural ecologists assume phenotypic plasticity, and a human ability to assess payoffs and/or learn from others the best alternative under a given set of ecological and social circumstances."

One critique of human behavioral ecology is that the formal models that form the foundation of this approach assume that people have the perfect knowledge of resources (and the time it takes to procure and process them) and that the ultimate goal is to efficiently use time and energy while maximizing reproductive fitness. However, in practice, people rarely meet these assumptions, and therefore critics would argue that human behavior is too complicated to formally model, or, as Kenneth E. Sassaman (2010:146) succinctly states, we are not "raccoons or blackbirds."

This critique is part of a larger debate in the social sciences over whether humans behave rationally, and/or do they behave rationally enough for social scientists, including archaeologists, to model their behavior. The academic debate over human economic rationality extends back to John Stuart Mill's (1844) assumption—and its subsequent critiques (Persky 1995)—that all humans seek to acquire wealth. This, admittedly, is a huge topic. However, on one hand, like Sassaman (2004, 2010), I would agree that hunter-gatherers and early farmers are not preprogrammed robots on a mission to economize for calories, and I have concerns over the use of terms like *hunter-gather* or *band-level society* and the neo-evolutionary baggage that these terms bring with them. Historical context matters, which makes reducing human behavior to a series of "if environment is x, then the expected behavior is y" statements extraordinarily difficult. Moreover, relishing the opportunity to explore the variation caused by historical context is something that should be embraced (e.g., Sassaman 2010).

On the other hand, Robert L. Bettinger (2009), Robert L. Kelly (2013), and Kennett and Winterhalder (2006) defend the use of formal models as a point of departure for examining variation in the archaeological record. In particular, I find Kelly's (2013) argument persuasive, especially with how he deals with the rationality conundrum. Kelly argues that human behavioral ecology now has a well-developed set of formal models for predicting how people *should* behave *if* their primary concerns were economizing for time, energy, and reproduction. He acknowledges that while people often fail to behave as preprogrammed robots singularly focused on acquiring calories and generating viable offspring, we can use the formal models from human behavioral ecology to ask, what if they were? What should they do? In economics, this caricature has a name: Homo economicus (Persky 1995). In statistics, this is called a null model, or the statement that one attempts to gather data to falsify. For archaeology, the formal models of human behavioral ecology allow us to create a baseline, or frame of reference (i.e., Binford 2001), for asking, are prehistoric people acting any different from how you would expect Homo economicus to behave? And if so, how?

Kelly (2013) contends that the expected behavior from behavioral ecology actually provides pretty reasonable null models because, despite the great variability in historical context in space and time, there's a general tendency for people to behave in ways that maximize their time and energy. The reason, he and other proponents of a "weak sociobiological thesis" would argue is that people tend to "select behaviors from a range of variants whose net effect, on average, in a given sociological and ecological context, is to maximize individual reproductive fitness." Furthermore, he contends that the proponents of human behavioral ecology, "assume there is plenty of wiggle room because of several factors, a notable one being human culture" (Kelly 2013:33).

Furthermore, based on the broad literature generated by proponents of human behavioral ecology and its cousin in academia, behavioral economics, it turns out that using Homo economicus is a pretty robust null model, even if actually finding an individual who behaves like Homo economicus all day everyday is highly unlikely. Or, as Box and Draper (1987:424) state, "Essentially, all models are wrong, but some are useful." Kelly (2013), other proponents of human behavioral ecology (Bettinger 2009; Kennett and Winterhalder 2006), and behavioral economists (Thaler 2016) would argue that the formal models that predict what Homo economicus would do are not likely to precisely predict human behavior, but they are nonetheless very, very useful.

On the other hand, if archaeologists are able to discern that people are not economizing for time, calories, or reproductive fitness, is that the end of the line for the usefulness of the formal models of behavioral ecology? To counter this critique, I would argue that while human behavior is complex, it is not random. A sizable amount of literature argues that human beings are so adept at recognizing patterns that we are cognitively incapable of consciously creating truly random distributions (e.g., Bennett 1998; Figurska et al. 2008; Mlodinow 2008). As a result, while culture surely does vary outside the realm of strict energetics and reproductive fitness, it is safe to say that humans carry out tasks that are goal-driven to some purpose, whatever that may be. Individuals operating outside predictions of the formal models of human behavioral ecology do not undermine the usefulness of optimality models. Instead, they indicate situations where people are behaving in a way that is not explainable as directly optimizing for caloric intake, and optimality models provide a point of departure for developing alternative explanations for patterns in the archaeological record.

For example, Bliege Bird and Smith (2005:221) argued, in their application of signaling theory to human behavior, that individuals often engage in "wasteful" behavior, or behavior that does not directly and efficiently contribute to subsistence or reproduction. In other words, at face value, they don't appear to be behaving like Homo economicus. Yet, when individuals display unconditional generosity, engage in wasteful subsistence behavior, and/or create artistic or craft traditions, they "might signal particular hidden attributes, provide benefits to both signaler and observers, and meet the conditions for honest communication." These behaviors are not random and they are not really wasteful. They are conveying a signal, or information, to other individuals.

As an archaeological example that operationalizes signaling theory, McGuire and Hildebrandt (2005) found that Middle Archaic hunter-gatherers in the Great Basin exploited ungulates to such a degree that it was above and beyond the expectations generated by optimality models—or they were acquiring much more than Homo economicus would if the goal was simply to acquire enough calories for subsistence. Instead, McGuire and Hildebrandt hypothesized that the prehistoric hunters had another goal for which they were economizing: prestige.

As another example, Thomas H. McGovern's (1994) analysis of the failure of the Norse colonies on Greenland is cited as a case study for historical ecology (e.g., Crumley 1994); it is also an instance of using the formal models of behavioral ecology as a point of departure for examining the archaeological record. While climate change is often cited as the prime mover for abandonment of the Norse colonies, McGovern argued that it is more interesting to ask why they were not more resilient in the face of climate change. The reason, he argued, was that the descendants of the initial colonizers of Greenland, as well as the Catholic Church, controlled a disproportionate amount of the most productive land, and they were unwilling to relinquish tracts to settlers living in more marginal areas. Second, while obvious examples from nearby indigenous groups illustrated how to survive during a climatic downturn, the Norse colonists stubbornly clung to their European, agrarian lifestyle.

Consequently, the very processes by which the Norse colonized Greenland, as well as their refusal to adopt alternative subsistence strategies, inhibited their ability to adapt and made their continued prospects for settlement unsustainable. While the old adage "hindsight is 20/20" applies here, McGovern's argument for what the Norse should have done is built on the back of models from human behavioral ecology. In this case, McGovern argued that the Norse should have broadened their diet breadth and reapportioned land to make agricultural production and distribution more efficient and equitable. However, this did not happen, and the Norse built larger churches, which led McGovern to generate an alternative hypothesis similar to the one posed by McGuire and Hildebrandt (2005) about hunter-gatherers in the Great Basin during the mid-Holocene: the goal is not to acquire calories efficiently but rather to acquire prestige. However, in the case of the Norse in Greenland, McGovern hypothesized that the retention of power by the elite is the most parsimonious explanation for their behavior.

A second issue with utilizing optimality models that is specific to archaeology is that while behavioral ecology is clearly not ahistorical, the studies that utilize these models do have a tendency to focus on decisions over relatively short time frames. Formal models such as the marginal value theorem (Charnov 1976) and the diet breadth model (MacArthur and Pianka 1966) are designed to make predictions regarding the instantaneous decisions of specific individuals with well-defined parameters. Archaeological datasets, on the other hand, are more likely to be aggregations of many decisions made by many people spanning multiple generations. Michael A. Jochim (1991:308) argued that this scalar mismatch is most obvious when researchers attempt to map idealized settlement models derived from optimal foraging theory and ethnographic observations onto the archaeological record. This temporal disjunction could also be viewed as one of the potential pitfalls of extrapolating ethnographic data to the archaeological record that H. Martin Wobst (1978) cautioned against in his classic article on "the Tyranny of the Ethnographic Record."

Fundamentally, the mismatch between formal models that predict what an individual will do within well-defined parameters and the archaeological record that is an aggregate of decisions made by many people over long periods of time has a name: the palimpsest problem. Or, as Binford (1981:197) framed it, "Rates of deposition are much slower than the rapid sequencing of events which characterizes the lives of living peoples, even under the best of circumstances, the archaeological record represents a palimpsest of derivatives from many separate episodes." Or, to rephrase this, it is very difficult to use observations of people in the present as a Rosetta stone to directly interpret patterns in the archaeological record.

While Binford was specifically addressing attempts to directly use the ethnographic record to interpret the archaeological record, it is relevant to the formal models of behavioral ecology because at the time he was writing about the palimpsest problem, ethnographic research was geared toward applying the models of human behavioral ecology to groups in a wide variety of environmental contexts (Kelly 2013). What subsequently ensued was a flurry of theory building on how to make the jump from using observations from the present (i.e., ethnographies and ethnoarchaeological data) to make inferences about the past (i.e., patterns in the archaeological record).

Overcoming the palimpsest problem was a primary concern of processual archaeologists (Shott 1998). Binford (1977) adopted the term *middle range theory* to describe the method and theory that archaeologists used to derive inferences regarding past human behavior from the archaeological record. While some question whether his use of the term *middle range theory* was appropriate (Raab and Goodyear 1984; Schiffer 1988), what grew from the research of Binford (1977, 2001), Michael Schiffer (1976, 1988), Clarke (1968) and others was a body of literature that falls under the labels of "site formation theory" (Shott 1998), "taphonomy" (Lyman 1994), and the "organization of technology" (Nelson 1991). I would argue that archaeologists developed a robust canon that allows us to translate patterns in the archaeological record into reasonable inferences about what people were doing in the past.

Of particular relevance here is the body of research that grew from Binford's research, which Margaret Nelson (1991) described as studies of the organization of technology. As this body of theory developed, its practitioners sought to increase the clarity of terminology (Kelly 1992; Shott 1996) and replace narrative models with formal ones (Kuhn 1990, 1994), which included borrowing heavily from human behavioral ecology (e.g., Beck et al. 2002; Kuhn 2004; Kuhn and Miller 2015; Shott 2015; Surovell 2009). In other words, archaeologists have come to grips with the ways in which people create time-averaged archaeological assemblages, how that record can be biased by both cultural and natural processes, and then how those assemblages can be used to address primarily economic questions, particularly as they relate to topics like subsistence and mobility.

As a relevant example, Mary C. Stiner (2001) analyzed variation in subsistence strategies between Neanderthals and modern humans in the eastern Mediterranean, which is built upon the analysis of many archaeological assemblages, or time-averaged remains of many individual decisions over thousands of years. She found that Neanderthals focused on prime-age adult ungulates and smaller game that were slow and easy to catch and argued that the consistent use of such high-return prey is predicted by the diet breadth model. Neanderthals were able to do this consistently because they appeared to have relatively low population densities. Later Upper Paleolithic humans, on the other hand, quickly depressed the population of slow-moving small game (which also have slower reproductive rates) and larger game species. They then incorporated fast-moving small game, a wider age profile of ungulates, and costlier (but nutritious) vegetable foods into their diet. This different subsistence signature is most likely due to greater rates of sustained population growth, which led to the depletion of local game populations and a greater investment in lower-ranked species.

Like McGuire and Hildebrandt (2005) and McGovern (1994), Stiner (2001) created a null model based on expectations from formal models of human behavioral ecology. In this case, she argued that, based on the diet breadth model, the array and frequency of species targeted as prey should vary with climate change. However, she found diet breadth continued to broaden *in spite of* climate change, leading her to conclude that an alternative hypothesis may be contributing to changes in diet over time: population pressure.

In this sense, one could argue that Stiner's research is an example of prehistoric macroeconomics, or more simply, a quantitative approach to deep history (e.g., Stiner and Feeley-Harnick 2011). Yet that deep history of subsistence change was built upon the careful examination of a large number of assemblages—from taphonomic biases that lead to the differential preservation of bone, the expected frequencies of species in time-averaged assemblages, and the effects of scavenging and the introduction of bone into archaeological deposits by other species (Stiner 1990a, 1990b, 1991a, 1991b, 1993, 1994; Stiner et al. 1995, 1999, 2000, 2001). In other words, Stiner (2001) had to employ over forty years of method and theory in archaeology geared toward untangling the various processes that generate, alter, destroy, or preserve archaeological sites before she could reconstruct that deep history. Given the success of Stiner's approach of using the models of human behavioral ecology to explore trends in subsistence in the millennia leading up to the origins of agriculture in the eastern Mediterranean, I find her approach a useful guide for examining another region: eastern North America.

Applying a Behavioral Ecological Approach to the Origins of Agriculture in Eastern North America

Conducting detailed analyses of prehistoric subsistence decisions is notoriously difficult because floral and faunal datasets are few and far between, especially in eastern North America, where the preservation of organic materials is biased toward shell midden and rockshelter sites (Styles and Klippel 1996). In order to test the various competing models for the transition to agriculture, we must look to other datasets. For better or worse, stone tools are the best preserved and best represented artifact class with which to observe changes in human culture over long periods of time.

Fortunately, the lower Mid-South, and the lower Tennessee and Duck River valleys in particular, is known for its abundant and diverse lithic assemblages from the Paleoindian and Archaic periods (Amick 1987; Anderson et al. 2010; Broster and Norton 1996; Broster et al. 2013; Lewis and Kneberg 1959). The Tennessee and Duck River drainages in central Tennessee contain multiple chert-bearing limestone formations that provide ready sources of raw material for stone tools (Amick 1987). Many have noted that the proximity to raw material sources contributed to high rates of artifact discard (Andrefsky 1994; Beck et al. 2002; Ingbar 1994; Kuhn 2004; Surovell 2009). In central Tennessee, this tendency is reflected in the abundance and ubiquity of early archaeological sites and artifacts recorded in private collections (Broster and Norton 1996; Broster et al. 2013). Moreover, the abundant sources of lithic raw material may have contributed to the extraordinarily high number of Paleoindian and Early Archaic bifaces recovered in Tennessee. Benton and Humphreys counties have some of the highest densities of Paleoindian period artifacts, as well as one of the most abundant and continuous records of stone tool technology for the Paleoindian and Archaic periods in eastern North America (Anderson et al. 2010; Prasciunas 2011).

More importantly, these two counties are near the geographic center of the distribution of sites with the earliest documented domesticated plant remains in eastern North America. They are located at the mouth of the Duck River, which has been extensively surveyed and contains the Hayes site, which has yielded the earliest documented domesticated sunflower seeds (Crites 1987; Smith and Yarnell 2009). The Duck River traverses multiple physiographic provinces, and it is located in the Mid-South, a region that contains the densest archaeological record of the early colonists of North America (Anderson and Gillam 2000; Meltzer 2009). This region is also well known archaeologically for subsequent resource intensification and plant domestication (Smith and Yarnell 2009).

The state of Tennessee also has an active Paleoindian projectile point survey (Broster and Norton 1996; Broster et al. 2013) and site file records that are good sources of spatial information about patterns of land use at different periods (Anderson 1996; Prasciunas 2011). Analysis of the distributions of sites and temporally diagnostic artifact forms allows for the identification of areas that were most intensively utilized by the initial occupants of these drainages and helps establish when humans expanded into (or abandoned) more marginal areas. These two sources of information combined enable me to evaluate variation in demography and hunting returns in the time periods preceding the inception of domesticated plants as a means to formally test widely cited models for the origins of agriculture.

Research Questions

This study examines the possible roles of population packing, intergroup competition, and differential mobility in regard to the origins of domestication and agriculture in eastern North America. More specifically, I assess Smith and Yarnell's (2009:6565) assertion that "there does not appear to be much, if any, evidence that landscape-packing and resource competition played a causal role in either the initial domestication of eastern seed plants or their coalescence into an initial crop complex."

The first part of the study involves a detailed evaluation of Paleoindian and Archaic period biface assemblages to make inferences regarding changes in technological organization use in the lower Tennessee River valley using economic approaches derived from (or related to) behavioral ecology (e.g., Kuhn 1995, 2004; Surovell 2009). Then, in order to examine demographic trends, I utilize the ideal free distribution (Kennett et al. 2006; McClure et al. 2006; Sutherland 1996), a formal economic model sometimes used by behavioral ecologists to predict when a species should exploit a new habitat. This model is used to interpret changes in the spatial distribution of temporally diagnostic artifacts and archaeological sites in the lower Tennessee and Duck River drainages.

In Chapter 2, I discuss the environmental setting and culture history in the southeastern United States, which includes a compilation of radiocarbon dates associated with temporally diagnostic bifaces. In Chapter 3, I provide an overview of recent approaches to the study of biface technological organization and outline an analytical method from behavioral ecology (e.g., Kuhn and Miller 2015) as a means to make inferences about prehistoric residential mobility and subsistence. I then use two case studies—Puntutjarpa Shelter in Australia (Gould 1977) and Gatecliff Shelter in Nevada (Thomas 1983)—to demonstrate how patterns in lithic technological organization relate to prey size and hunting returns.

In Chapter 4, I apply these insights to a sample of sites selected from Benton and Humphreys counties that spans the Early Paleoindian through Late Archaic periods (>13,000–3000 cal BP), which is one of the only areas in eastern North America with an abundant and continuous archaeological record—from initial colonization to the domestication of indigenous plants. In Chapter 5, I discuss previous attempts to model prehistoric demography in eastern North America, with a particular emphasis on studies using the distribution of radiocarbon dates and archaeological sites. I then use the ideal free distribution (e.g., Fretwell and Lucas 1969; Sutherland 1996) to interpret the distribution of archaeological sites through time in the Lower Tennessee and Duck River valleys in order to assess demographic trends in the Mid-South.

In the final chapter, I argue that the origins of plant domestication came about within the context of a millennial-scale boom-bust cycle that has its roots in the Late Pleistocene and that culminated in the mid-Holocene. More specifically, warming climate caused a significant peak in the availability of shellfish, oak, hickory, and deer, which generated a tipping point during the Middle Archaic period where hunter-gatherer groups narrowed their focus on these resources. After this boom ended, some groups shifted their focus to other plant resources that they could intensively exploit in the same manner as oak and hickory, which includes the suite of plants that were subsequently domesticated. This bust is likely due to the combined effects of an increasing population and declining returns from hunting, which is evident in my analysis of biface technological organization and site distributions from the lower Tennessee and Duck River valleys. Consequently, this study provides a nuanced understanding of demography and environmental effects in the time periods that preceded the appearance of domesticated plants.

CHAPTER 2

Environmental and Chronological Building Blocks

Introduction

Testing hypotheses for why people in eastern North America domesticated plants requires an in-depth understanding of the regional culture history, major trends in climate and regional proxy records for past environments, and artifacts that allow archaeologists to bridge the latter with the former. These steps are the proverbial building blocks that allow archaeologists to construct hypotheses about the relationship between climate and culture similar to Stiner's (2001) research in the eastern Mediterranean. In other words, is variation in the archaeological record a response to changes in the environment? To build the same null model for this study, I provide a broad overview of the culture history of eastern North America and how it relates to global trends in climate and how those changes are reflected in local environmental records. Finally, I argue that bifaces are an artifact class by which we can construct a fairly accurate chronology for the lower Tennessee and Duck River valleys.

The Culture History of Eastern North America

The question of how long people have been in North America has been the focus of considerable research since the middle of the nineteenth century (Haven 1856; Meltzer 2009). In eastern North America, this debate has revolved around two interrelated topics. First, European colonists were intrigued by the many earthen mounds dotting the landscape (Anderson and Sassaman 2012:8–9). While they attributed their construction to a variety of groups—from the Ten Lost Tribes of Israel to the mysterious race of Mound Builders—Cyrus Thomas (1894) eventually demonstrated that the ancestors of contemporary Native Americans constructed these mounds (Anderson and Sassaman 2012:15–16).

From this point forward, researchers shifted their emphasis to determining how long Native Americans had been in North America, driven in part by discoveries of Paleolithic archaeological sites in Europe and elsewhere (Meltzer 2009:68–69). This debate was amplified with the announcement by Charles C. Abbott (1877, 1889) that "Paleolithic" artifacts had been discovered in New Jersey. William Henry Holmes (1890) later showed that the simple fact that these artifacts were crudely made did not mean that they were very old. Later, Holmes and Aleš Hrdlička became widely renowned critics of a deep antiquity of people in North America (Meltzer 2009:68–79). It was not until the discovery at Folsom, New Mexico, where stone projectile points (i.e., Folsom points) were found in association with extinct species of bison, that a Pleistocene-aged human presence in the Americas was established (Meltzer 2006). In 1933 Edgar B. Howard and John Cotter began excavations at a gravel quarry between Clovis and Portales, New Mexico, where fluted points were reportedly found in association with the remains of extinct species of bison and mammoth (Boldurian and Cotter 1999). Much later, Elias H. Sellards and Glen Evans identified the stratigraphic distinction between Clovis and Folsom, which at that time made Clovis the oldest documented archaeological culture in North America (Sellards 1952). While many sites are purportedly older than Clovis in North America (Meltzer 2009:95–135), a continuous human presence in North America dates to at least 13,250 cal BP (Waters and Stafford 2007:1123).[1]

There is an extensive record of surface finds in eastern North America yet very few stratified sites containing Pleistocene-aged deposits and radiocarbon datable material (Dunnell 1990:13). Instead, most fluted bifaces, the key temporally diagnostic artifact from the earliest occupations across the continent, come from shallow sites, usually plowed fields. Furthermore, professional archaeologists do not discover most of these artifacts, which instead reside in private collections (Goodyear 1999:433).

David J. Meltzer (1988) interpreted this pattern of scattered isolated finds and few buried sites as evidence of extreme mobility, whereby small groups moved rapidly across the landscape, leaving little aside from small sites and scattered, isolated bifaces. Robert C. Dunnell (1990:13) argued instead that the ubiquity of shallowly buried sites is most likely due to a broad scale geomorphological bias. He observed that compared to other areas in North America, the southeastern United States is situated on a much older landscape, with many upland areas receiving little to no sedimentation since the arrival of people. Additionally, the warm, mesic climate of the region promotes the decay of materials that can be radiocarbon dated. That these factors inhibit the preservation of organic materials is reflected in recent continental-scale databases of radiocarbon dates, which contain comparatively few dates in the southeastern United States compared to other parts of North America (Buchanan et al. 2008; Miller and Gingerich 2013a; Waters and Stafford 2007). As a result, archaeologists studying the Pleistocene archaeological record in eastern North America, and especially the southeastern United States, rely heavily on a handful of sites in the region, supplemented with well-dated sites outside the region, for creating a culture-historical framework (Miller and Gingerich 2013b).

While there are a limited number of stratified Pleistocene sites in eastern North America, the Holocene-aged record has benefited greatly from extensive and systematic archaeological research over the last 80 years. This boom was in part spurred by the creation and funding of the Works Progress Administration (WPA) in the 1930s, which employed many individuals in archaeological projects across North America in advance of reservoir construction and other public works projects to combat unemployment in the wake of the Great Depression (Means 2013). In the southeastern United States, additional public works projects commissioned by the Tennessee Valley Authority (TVA), the Army Corps of Engineers, and other government agencies continued this tradition of funding archaeological research (Anderson and Sassaman 2012:19, 22–30).

As a result, the culture-historical sequence for eastern North America began to take form in the middle of the twentieth century. In particular, Joffre Coe's stratigraphic excavations at the Hardaway and Doershock sites along the Pee Dee River in North Carolina (Coe 1964; Daniel 2001), along with research in Tennessee (Lewis and Kneberg 1959) and northern Alabama (Cambron and Hulse 1969; Webb 1939; Webb and DeJarnette 1942), provided much of the critical foundation upon which the contemporary culture-historical framework is based. With the addition of sites such as Koster and Napoleon Hollow in Illinois (Brown and Vierra 1983); St. Albans in West Virginia (Broyles 1966); and Icehouse Bottom, Bacon Farm, and Rose Island in Tennessee (Chapman 1976; Kimball 1996); the culture-historical sequence for eastern North America was reinforced by further stratigraphic excavations as well as a growing number of radiocarbon dates.

Formalizing the Culture-Historical Sequence

In the early days of radiocarbon dating (e.g., Libby 1952), many of the first archaeological radiocarbon dates in eastern North America were produced by the laboratory at the University of Michigan at the behest of James Griffin, the director of the Museum of Anthropology. These include the dates from St. Albans in West Virginia (Broyles 1966), Eva in Tennessee (Lewis and Lewis 1961), and Graham Cave in Missouri (Crane and Griffin 1968). These dates, along with Griffin's access to one of the premier collections of artifacts in North America, provided much of the underpinning for his broad culture-historical sequence for eastern North America (e.g., Griffin 1952, 1967).

Griffin (1952) initially divided the prehistory of North America into the Paleoindian, Early Archaic, Late Archaic, Early Woodland, Middle Woodland, and Mississippian periods based on variability in material culture, inferred social organization, and subsistence. He focused on putting the attributes of these divisions in relative order. In a later update of this chronology, Griffin (1967) elaborated on his culture-historical sequence and included absolute numerical date ranges for each of his periods. This general framework is still used by most archaeologists who work in eastern North America (Anderson and Sassaman 2012:5; Table 2.1).

Smith (1986) contends that subsequent attempts to update this framework have often settled on either a "cultural" or "natural" approach. For example, Vincus P. Steponaitis (1986) divided the chronological framework along the same lines as Griffin (1967). Aside from a brief discussion of the environmental changes during the Late Pleistocene and Early Holocene, he focused almost solely on changes in material culture. On the other hand, Smith (1986) spends a significant amount of space in his

TABLE 2.1. James Griffin's culture history for eastern North America.

1952	1967
Paleo-Indian (12,000–8,000 years ago) Small groups hunting bison and mammoths, but likely utilized local flora and fauna in eastern North America	Paleo-Indian (9,500–8,000 BC) Small hunter-gatherer groups, probably responsible for hunting large mammals to extinction
Early Archaic (no date range) Patrilocal bands, mortars and pestles, hot rock cooking, shell heaps	Early Archaic (~8,000–6,000 BC) Some continuity with earlier groups, but thinly scattered, with more diverse subsistence
Late Archaic (no date range) Soapstone, bannerstones, steatite containers, copper, shell mounds, more elaborate bone tool technology and burials	Middle Archaic (6,000–4,000 BC) More ground stone and polished stone tools, elaborate bone technology, shell utilization
Early Woodland (no date range) Economy the same as Late Archaic but now includes burial mounds and more projectile point diversity; introduction of pottery; the appearance of small villages	Late Archaic (4,000–1,000 BC) "Time of considerable population growth, clear regional adaptation, and exchange of raw material" (178) Large shell middens, copper production, early pottery (fiber tempered)
Middle Woodland (no date range) Hopewell, large ceremonial centers	Early Woodland (1,000–200 BC) Adena complex, burial mounds, small villages, early agriculture
Mississippian (no date range) Pyramidal mounds, fortified villages, corn agriculture, matrilineal and matrilocal groups; continuity with groups observed by the first European explorers	Middle Woodland (200 BC–AD 400) Hopewell ceremonial centers and exchange networks
	Late Woodland (AD 400–AD 1,000) Introduction of maize agriculture and temple mounds
	Mississippian (AD 1000–European contact) Inception and spread of the Southeastern Ceremonial Complex

overview discussing major changes in vegetation (e.g., Delcourt et al. 1983), which was subdivided based on broad-scale variation in climate.

The publication of more detailed climatic reconstructions motivated David G. Anderson (2001) to discuss the broad impact of climate and culture in the southeastern United States. Perhaps the most notable study in this regard is the Greenland Ice Sheet Project 2 (GISP2), which contains a record of past temperatures and atmospheric conditions for the last 110,000 years (Alley 2000). From the GISP2 data, Richard B. Alley (2000) argues for abrupt changes in climate—most notably, the Younger Dryas cooling event (12,900–11,700 cal BP). While Anderson (2001:155) hypothesized the potential role of the Younger Dryas on early populations in North America, in the ensuing decade there has been considerable debate on what archaeologically perceptible impact, if any, the Younger Dryas may have had on people, especially beyond eastern North America and western Europe (Meltzer and Holliday 2010; Strauss and Goebel 2011).

Anderson (2001) designates the Paleoindian period as the time period coinciding with the initial appearance of people in North America and spanning the Younger Dryas (>13,450–11,450 cal BP/11,500–10,000 ^{14}C yr BP; Figure 2.1a; Table 2.2). Like Smith (1986), Anderson (2001) also sets the beginning of the Archaic period—which is further subdivided into the Early (11,450–8,900 cal BP/10,000–8,000 ^{14}C yr BP), Middle (8,900–5,700 cal BP/8,000–5,000 ^{14}C yr BP), and Late (5,700–3,200 cal BP/5,000–3,000 ^{14}C yr BP) periods—as the beginning of the Holocene.

These subdivisions are based on a loose correlation between climate and cultural change, with the Early Archaic period encompassing the most dramatic increase in temperature and sea level rise beginning at the end of the Younger Dryas. The Middle Archaic coincides with a time period of warmer, drier conditions in the mid-Holocene that has been labeled as the altithermal, hypsithermal, or mid-Holocene climatic optimum (Anderson et al. 2007). The Late Archaic occurs with the end of the warmer, drier mid-Holocene conditions and also coincides with the fluorescence of the Poverty Point site in Louisiana and regional trade networks (Anderson 2001:161–163).

The Woodland period is marked by the widespread adoption of pottery technology, the incorporation of domesticated plants into a full-fledged gardening complex, a collapse and reorganization of long-distance trade networks, and the emergence of the Adena and Hopewellian mound and earthwork complexes across eastern North America. The Mississippian period is characterized by the widespread adoption of maize agriculture, the appearance of shell-tempered pottery and wall-trenched houses, and the spread of the Southeastern Ceremonial Complex (Anderson 2001:163, 165–166).

TABLE 2.2. The current culture-historical sequence for eastern North America (adapted from Anderson and Sassaman 2012:5).

Calendar Dates (approximate)	Period	Culture Complex
>11,050 BC	Early Paleoindian	Pre-Clovis
11,050–10,950 BC	Middle Paleoindian	Clovis and other fluted point types
10,950–9,950 BC	Late Paleoindian	Dalton and other unfluted lanceolates
9,950 BC–6,900 BC	Early Archaic	Early Side-Notched, Corner-Notched, and Bifurcate
6,900 BC–4,350 BC	Middle Archaic	Benton and Watson Brake
4,350–1,200 BC	Late Archaic	Stallings Island and Poverty Point
1,200 BC–300 BC	Early Woodland	Adena
300 BC–AD 225	Middle Woodland	Hopewell
AD 225–AD 930	Late Woodland	Coles Creek
AD 930–AD 1350	Mississippian	Mississippian

Correlating Culture History and Climate in the Mid-South

While the GISP2 climatic reconstructions are informative about changes in climate across the North Atlantic, and to a certain extent at a global level, other studies have attempted to examine variation in climate at much more regional scales. For example, Viau and colleagues (2006) used a large database of pollen records to reconstruct the mean July temperature for the Holocene in North America, with one of their sub-samples being the southeastern United States (Figure 2.1b). While their temperature reconstruction bears a striking resemblance to the temperature reconstruction for GISP2, many of the other regions in their sample do not. However, one noticeable difference between GISP2 and the temperature reconstruction of Viau and colleagues (2006) for the southeastern United States is a more pronounced peak in temperature that lasted several centuries around 7,000 cal BP, coinciding with the altithermal or hypsithermal.

These broad-scale changes in climate correspond to significant changes in vegetation across eastern North America. Delcourt and Delcourt (1983, 1985) have outlined the variation in vegetation from the Last Glacial Maximum (LGM) to the present (Figure 2.2). Their reconstructions are based on a large sample of pollen records, which they then used to extrapolate floral composition and generate distribution maps based

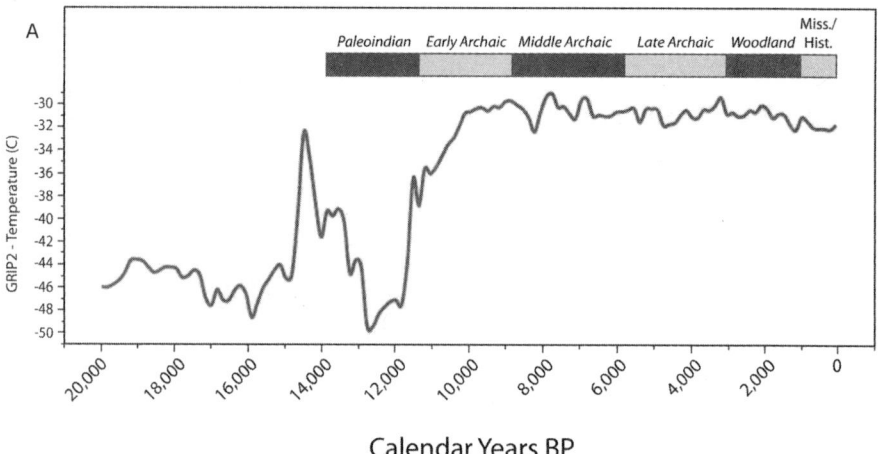

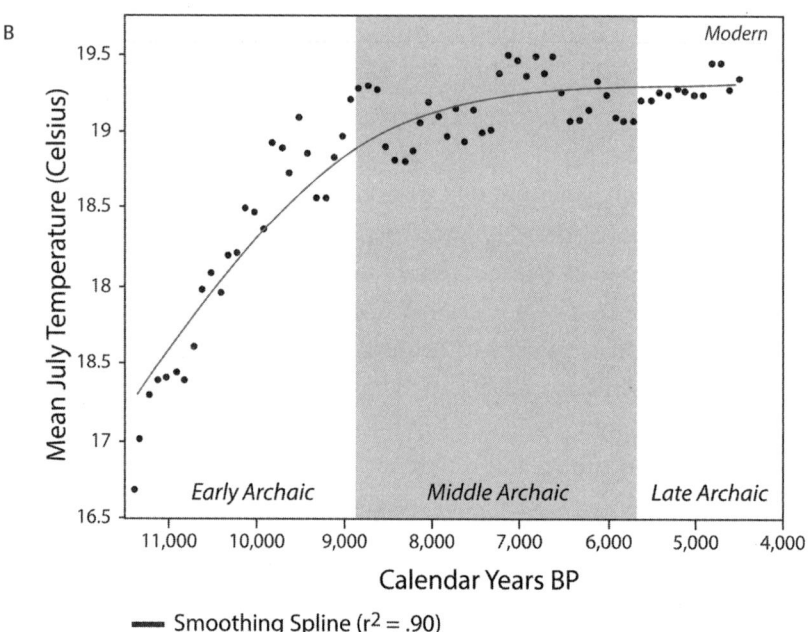

FIGURE 2.1. (*a*) Greenland Ice Sheet Project (GISP2) temperature reconstruction (Alley 2000) and the culture history periods for eastern North America (Anderson 2001). (*b*) Mean July temperature reconstruction for the southeastern United States (Viau et al. 2006) and the Archaic period culture-historical divisions for eastern North America (Anderson 2001)

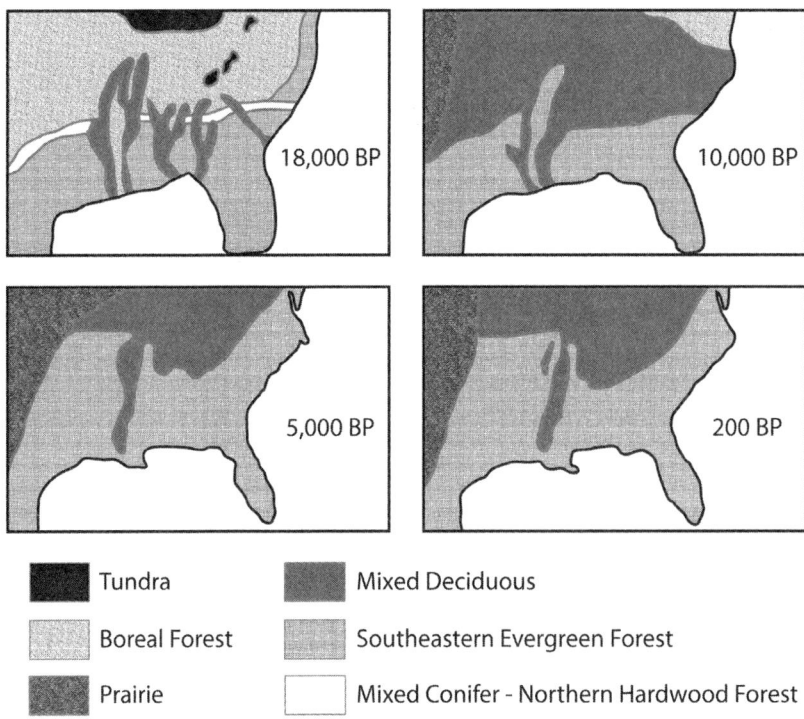

Tundra

Boreal Forest

Prairie

Mixed Deciduous

Southeastern Evergreen Forest

Mixed Conifer - Northern Hardwood Forest

FIGURE 2.2. Delcourt and Delcourt's (1985:16) vegetation reconstructions for eastern North America.

loosely on Nevin M. Fenneman's (1938) physiographic provinces. They argue that during the LGM (23,000–16,500 BP ^{14}C year BP/27,852–19,638 cal BP), much of eastern North America was covered in boreal forests composed of predominantly spruce (*Picea sp.*) but also fir (*Abies sp.*) and jack pine (*Pinus banksiana*). In some instances, these species occurred as much as 1,200 km south of their modern range in Manitoba and eastern Ontario. Also, during the LGM, mixed hardwood refugia have also been detected in the pollen records of the southeastern United States. These refugia contained a diverse mix of temperate species, including beech (*Fagus sp.*), maple (*Picea sp.*), walnut (*Juglans sp.*), hickory (*Carya sp.*), and oak (*Quercus sp.*). They occurred in a diversity of settings, including loess-capped uplands, ravines within sandy interfluves, and the irregular topography of karstic terrain.

During the Late Pleistocene (16,500 to 12,500 BP ^{14}C year BP/19,638–14,761 cal BP), there was a marked decrease in the prevalence of jack pine pollen relative to spruce and fir at several locations. Delcourt and Delcourt (1985) attribute this to the persistence of cool temperatures and an increase in precipitation. At the end of the Pleistocene, there was also an increase in the prevalence of oak and hickory pollen, which they interpreted as an expansion of mixed temperate species northward from their LGM refugia. In the Early Holocene (12,500–8,000 ^{14}C year BP/14,761–8,879 cal BP), Delcourt and Delcourt found that cool temperate mesic forests became more dominant, including mixed conifer and deciduous forests with no modern analog.

In the mid-Holocene (8,500–4,000 BP ^{14}C year BP/8,879–4,478 cal BP), Delcourt and Delcourt argued that increasing aridity and temperature resulted in the eastward expansion of prairie grasslands as well as lower species diversity and xeric conditions in central Tennessee. However, in the Ridge and Valley and Piedmont physiographic provinces, the species composition in places appears to reflect warmer and wetter conditions. On the Coastal Plain, the oak- and hickory-dominated forests were almost completely replaced by southern pine.

In the Late Holocene (4,000 ^{14}C year BP to present/4,478 cal BP), spruce and fir are restricted to the high altitudes and there was an expansion of oak-chestnut forests across the southern Appalachian Mountains. Shortleaf pine forests in Missouri and eastern Oklahoma expanded northward as the prairie grasslands retreated, most likely as the result of increasing precipitation. Also, Carolina bays begin to fill with peat and rising seas inundated coastal swamps. Importantly, Delcourt and Delcourt (1985) also note that, in many of their sample locations, there is increasing evidence for anthropogenic disturbance in the last 2,000 years, especially with the appearance of maize agriculture. Delcourt and Delcourt's (1985) reconstruction of past vegetation has since been updated, and Williams and colleagues (2004) have now made it possible to download the distribution of key pollen types for North America. For example, Figure 2.3 illustrates the northward expansion of oak-dominated forests from the Late Pleistocene to the present day. In the following sections, I discuss changing environmental conditions as reflected in more local proxy records found within the Highland Rim and Nashville Basin.

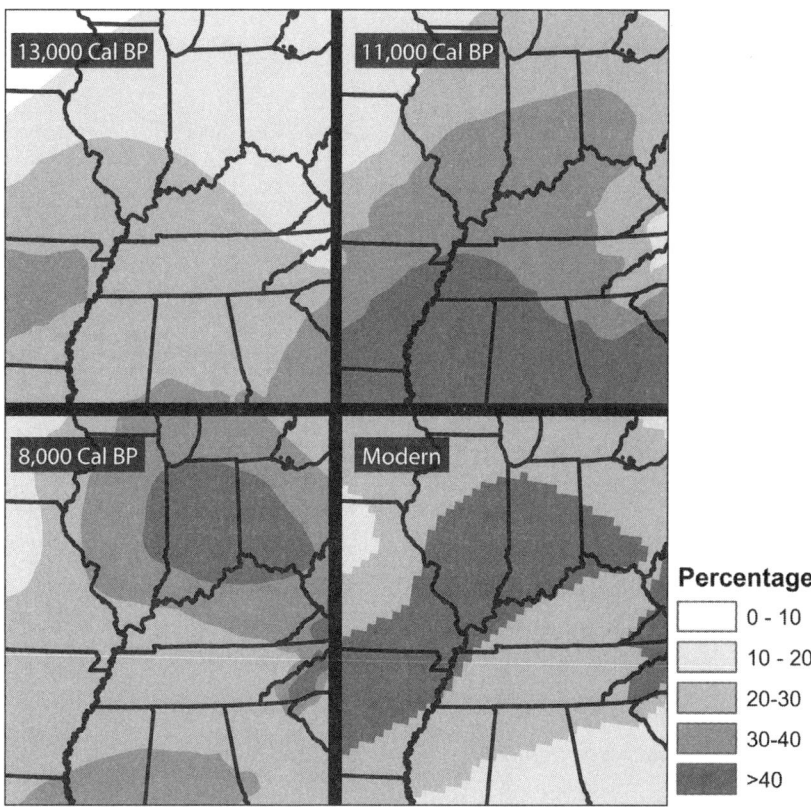

FIGURE 2.3. The distribution of oak pollen (*Quercus sp.*) in eastern North America based on interpolating count density from Williams et al. (2004).

Late Pleistocene and Holocene Environment in the Highland Rim and Nashville Basin

There are three classic paleoenvironmental studies for the lower Tennessee and Cumberland River drainages in central Tennessee (Figure 2.4). The first is Hazel R. Delcourt's (1979) analysis of pollen cores from Anderson Pond in White County, which contains a 25,000-year pollen record for the Eastern Highland Rim and was one of the sample localities for Delcourt and Delcourt's (1985) subsequent study of the major trends in vegetation, summarized above. At Anderson Pond, Delcourt (1979) inferred evidence for boreal forests during full glacial conditions from the presence of jack pine, spruce, and fir pollen. During the Terminal Pleistocene and Early Holocene, these species were replaced by a

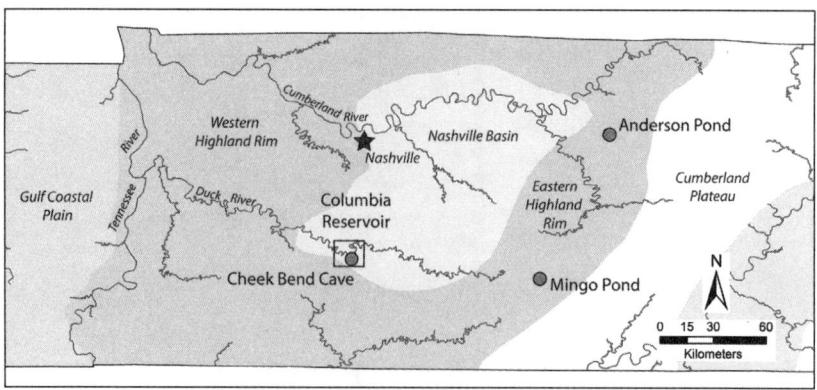

FIGURE 2.4. Map of major rivers and physiographic sections in central Tennessee with the locations of Anderson Pond, Mingo Pond, Cheek Bend Cave, and the Columbia Reservoir project highlighted.

more deciduous, mixed mesophytic forest composed of ash, ironwood, hickory, birch, butternut, willow, and elm, which were then followed by beech and sugar maple.

During the mid-Holocene, Delcourt argued, the composition of the forest indicates a major warming and drying trend that is reflected by an influx of oak, ash, hickory, swamp alder, and buttonbush. Moreover, the relative importance of mixed mesophytic species likely indicates warmer and dryer summer temperatures. However, after 5,000 ^{14}C years ago, this trend appears to have ameliorated, and the composition of the forest mirrored the modern forests present in the region. The results of this study also parallel two other analyses of pollen cores in the region—Mingo Pond (Delcourt 1979) and Jackson Pond (Wilkins et al. 1991).

Similar to Delcourt and Delcourt's (1983, 1985) observation regarding the change in forest composition, the pollen cores from both Jackson Pond and Anderson Pond show large increases in oak pollen during the mid-Holocene. After generating a correlation matrix for the pollen counts from Anderson Pond, I found that the influx of oak not only correlates strongly with hickory pollen but also positively correlates with 25 other species, which contrasts significantly with the coniferous species that dominated during the LGM (Figure 2.5). Consequently, forests became increasingly diverse during the Holocene in comparison to full glacial conditions.

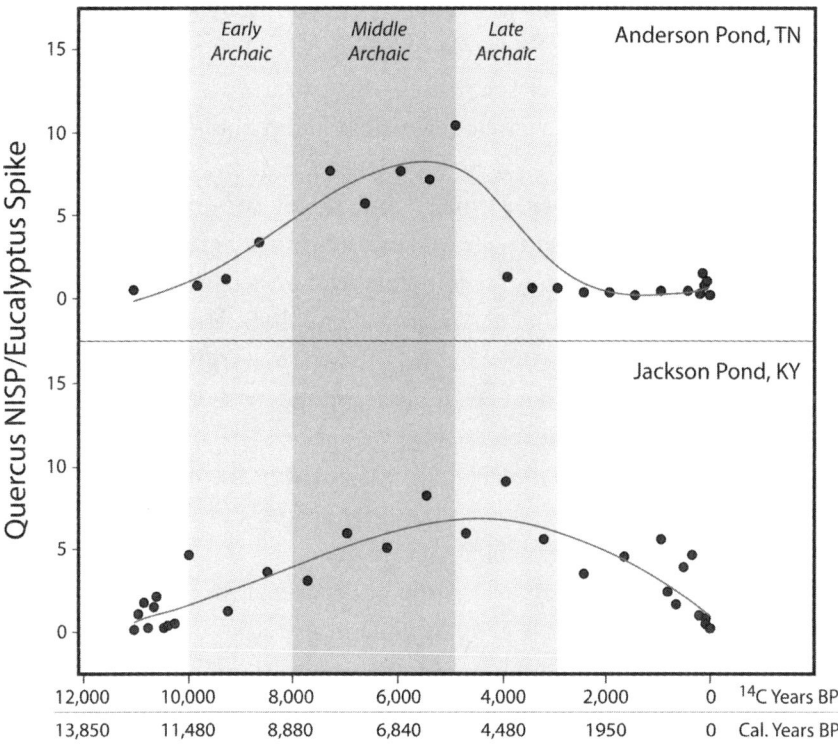

FIGURE 2.5. The distribution of oak (*Quercus sp.*) pollen during the Holocene for Anderson Pond, Tennessee (Delcourt 1979) and Jackson Pond, Kentucky (Wilkins et al. 1991) with the Archaic culture-historical periods (Anderson 2001) highlighted.

In the second major paleoenvironmental study, Klippel and Parmalee (1982) compared their analysis of the faunal remains from Cheek Bend Cave in Maury County, Tennessee, to the pollen records from Anderson Pond. This cave, excavated as part of the TVA-sponsored Columbia Reservoir project, consisted of eight strata that were Pleistocene to mid-Holocene in age. They argued that the faunal record in this cave—especially the small vertebrates (e.g., moles, shrews, mice, and rats)—represents the natural death assemblage since archaeological remains were present only at the uppermost strata, and there was no evidence for human consumption of these species. Instead, they argued that these species are valuable proxies for environmental change because they are sensitive to changes in temperature and precipitation.

Klippel and Parmalee found species in the lower strata of the cave that appear in modern boreal contexts today, and this likely represents the full glacial record for the region. This pattern was also apparent in a separate study of the avian fauna, which found boreal species, such as the northern hawk owl (*Surnia ulula*) and the boreal owl (*Aegolius funereus*) in the same strata (Parmalee and Klippel 1982). They noted a subsequent shift in the assemblage that may represent ameliorating climate congruent with Delcourt's (1979) interpretation of the Late Pleistocene and Early Holocene record at Anderson Pond. In Stratum IV, they found species adapted to more open habitats, which they argued was likely due to the expansion of upland cedar glades in the Nashville Basin, part of the expansion of more drought-tolerant species responding to warmer, drier conditions in the mid-Holocene. Finally, in the uppermost strata, they argue that the species reflect deciduous forest expansion at the expense of the cedar glade habitat.

In the third classic study, G. Robert Brackenridge (1984) conducted one of the few large-scale geomorphological studies of the region that, like the excavations at Cheek Bend Cave, were part of the TVA-sponsored Columbia Reservoir project. This study was based on an analysis of 18 backhoe trenches over a 22-km stretch of the Duck River in the Nashville Basin, representing a 30,000-year record of floodplain sedimentation, erosion, and stability that was dated using a combination of radiocarbon dating and temporally diagnostic archaeological remains (Figure 2.6). Brackenridge found that Late Pleistocene deposits are 5 m above the modern floodplain, and there is also evidence for severe bedrock and floodplain erosion likely dating to the end of the Pleistocene. This was followed by vertical and horizontal aggradation during the Early Holocene. By 7,200 ^{14}C year BP (8,000 cal BP), the floodplain stabilized and soil formation occurred, which he argued probably is related to a warmer, drier climate during the mid-Holocene. Around 6,200 ^{14}C year BP (7,084 cal BP), sedimentation again overtook soil development, which, he argued, indicates more humid conditions that were punctuated by two additional episodes of soil development that ended at 2,600 ^{14}C year BP (2,745 cal BP) and 150 ^{14}C year BP (143 cal BP).

These three records appear to reflect broader paleoclimatic trends in eastern North America (Figure 2.7). First, both the pollen record at

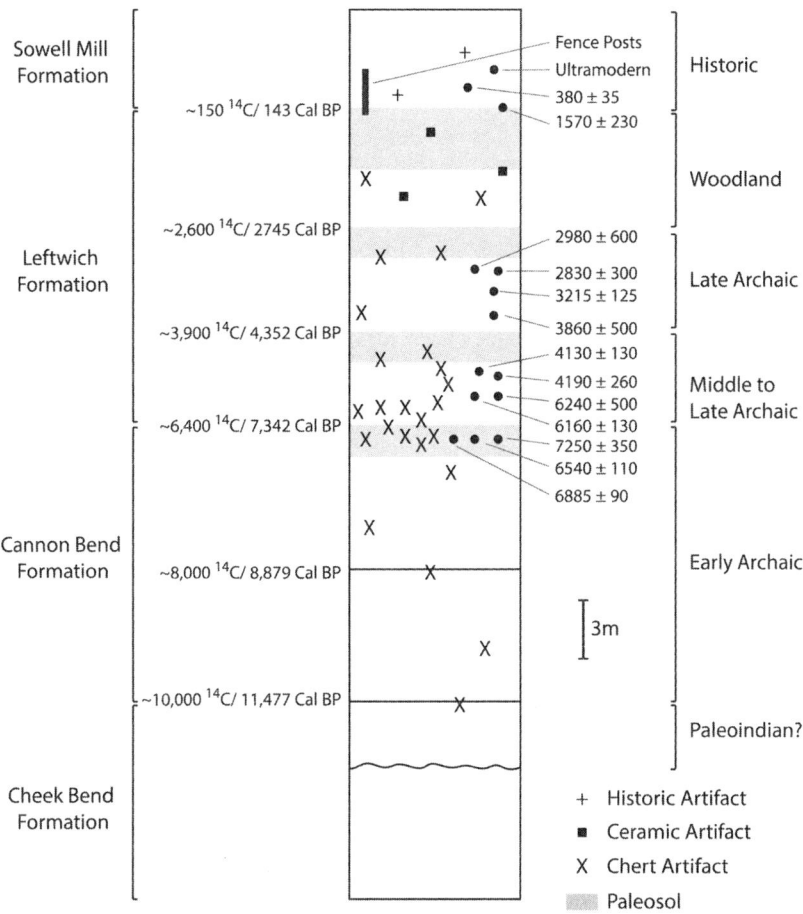

FIGURE 2.6. Composite geologic section for the Duck River (Brackenridge 1984:19).

Anderson Pond and the faunal record at Cheek Bend Cave provide evidence for boreal forests present at full glacial conditions, which then gave way to warmer, wetter conditions during the Terminal Pleistocene and Early Holocene. This transition is also evident in the geomorphological record, where a major erosional discontinuity may be related to the changeover in forest composition. The addition in precipitation increased the amount of water entering river systems, while the increase in vegetation decreased the sediment load entering the rivers. This would have provided the conditions for lateral erosion and downcutting of the

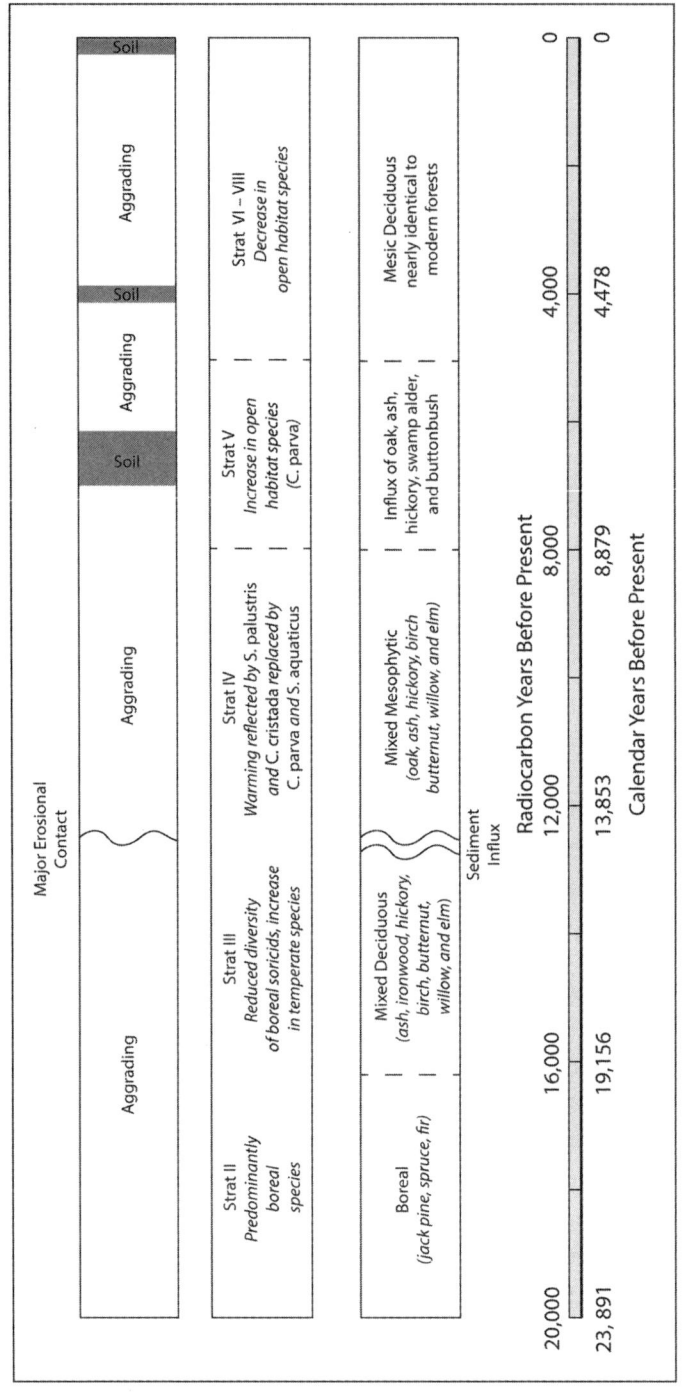

FIGURE 2.7. Composite paleoenvironmental reconstruction for the Nashville Basin and Eastern Highland Rim since the Last Glacial Maximum based on (top) Brackenridge's (1984) geomorphological study of the Duck River, (middle) Klippel and Parmalee's (1982) analysis of small vertebrate fauna at Check Bend Cave, and (bottom) Delcourt's (1979) vegetation reconstruction for Anderson Pond.

channel. On the coastal plain, this transition is reflected by a shift from braided to large, meandering rivers (Leigh 2008). While it is unclear if a similar transition occurred on the Duck River or any other drainage in the region, the erosional disconformity is most likely related to broad-scale changes in precipitation and land cover.

During the mid-Holocene, all three records are consistent with warmer and drier conditions. Interestingly, at Anderson Pond, the warmer/drier conditions are reflected by an influx of oak and hickory pollen, whereas Klippel and Parmalee argue that in the uplands of the Nash-ville Basin, cedar glades expanded. Both Delcourt (1979) and Klippel and Parmalee (1982) argue that this is indicative of less rainfall during the summer months. This interpretation is also consistent with the mid-Holocene soil observed by Brackenridge (1984). Warmer temperatures coupled with less precipitation would reduce sedimentation and foster soil development. Finally, at the end of the mid-Holocene, modern oak-hickory forests became established, which is reflected in both the pollen records at Anderson Pond and an increase in more closed-habitat species at Cheek Bend Cave, which Klippel and Parmalee interpret as evidence for the reduction in the distribution of upland cedar glades. During this time span, Brackenridge noted that the Duck River began aggrading again, with the exception of two periods of soil development that ended at 2,600 ^{14}C year BP (2,745 cal BP) and 150 ^{14}C year BP (143 cal BP).

Paleoindian and Archaic Period Culture History

In the southeastern United States, sites with radiocarbon datable components are rare, especially those dating to the Later Pleistocene and Early Holocene (Anderson 2005; Dunnell 1990; Goodyear 1999; Miller and Gingerich 2013a). Consequently, the culture-historical framework is based on a relatively small number of dated sites, and archaeologists use temporally diagnostic artifacts such as projectile points and pottery to cross-date sites elsewhere. However, the oldest dated pottery in eastern North America is Stallings Island fiber-tempered pottery, which dates to 3,500 cal BP (e.g., Sassaman et al. 2006). As a result, archaeologists have to rely on regional projectile point typologies for the entirety of the Paleoindian period and almost all of the Archaic period in instances where radiocarbon dates are not available.

Of course, using temporally diagnostic artifacts to date archaeological assemblages is not without its pitfalls. For example, it is still unclear as to why projectile point shapes change over time in such a way that their shape and other stylistic attributes can be used as temporal markers. While this could be related to changes in the prey or hafting technology (e.g., Buchanan et al. 2011), it could also reflect evolving communities of practice (e.g., Bettinger and Eerkens 1999; Thulman 2006) or random change operating on the parts of the projectile points that are not visible when hafted to a spearshaft (e.g., Hamilton and Buchanan 2009). Moreover, many of the sites producing the best associations between distinctive artifact types were excavated before the first major publications on site formation theory, and most dates were produced prior to the development of acid-base-acid pretreatment and the inception of the accelerator mass spectrometry (AMS) method for deriving more precise dates (Bronk Ramsey 2008; Miller and Gingerich 2013a). In order to generate a more refined chronology, Bissett and Miller (2017) conducted a Bayesian statistical analysis using a sample of 247 radiocarbon dates from 55 sites that have produced Paleoindian and Early Archaic bifaces (Figure 2.8).

For this study, I begin with the fluted point occupation for the Mid-South. This is generally thought to begin with the Clovis type, which is for the most part described as large parallel-sided lanceolate bifaces with slightly concave bases and single or multiple flutes that rarely extend more than a third of the body (Howard 1990; Justice 1995:17–21; Morrow 1995; Sellards 1952; Smallwood 2012) (Figure 2.9a). Bifaces meeting these criteria have been reported in almost every state in North America (Anderson et al. 2010; Miller et al. 2013). Secondary hallmarks of Clovis lithic technology include the presence of biface caches (Kilby 2008) and prismatic blades (Collins 1999). Based on a sample of sites from across North America, Waters and Stafford (2007:1123–1124) argue that the Aubrey site in Texas is likely invalid, and the earliest date range for the appearance of Clovis points should be ~11,050 ^{14}C yr BP (12,937 cal BP). On the other hand, Haynes and colleagues (2007) use a broader date range for Clovis (11,500–10,800 ^{14}C; 13,351–12,677 cal BP) in part because of C. Reid Ferring's (2012) defense of the dates at Aubrey. Bissett and Miller's (2017) model produced a comparable date range for Clovis.

After Clovis, there are four other point types that I divide into two groups. First, Gainey and Redstones are lanceolate points that are very

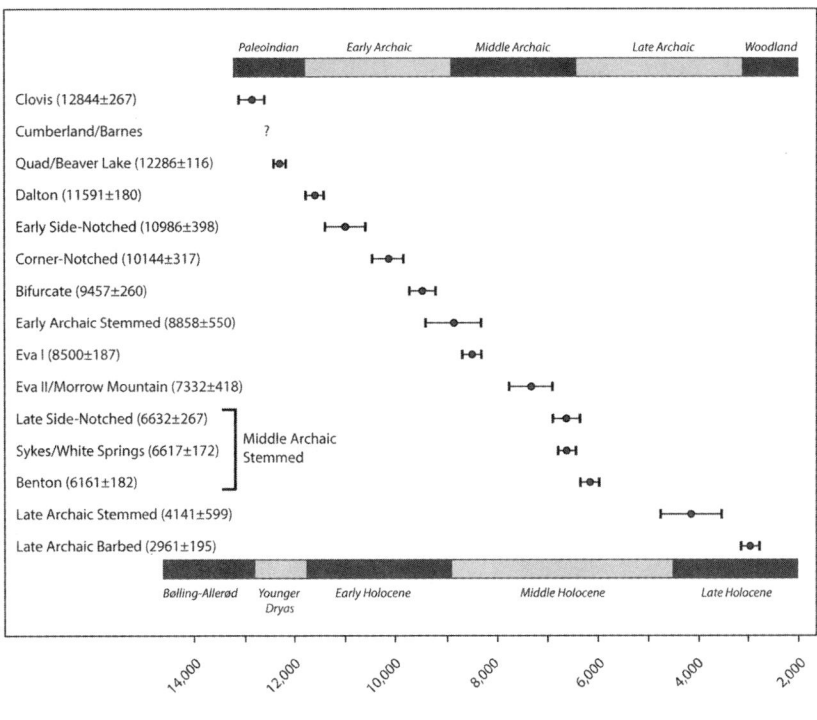

FIGURE 2.8. Date ranges (mean and one standard deviation) for selected projectile point types in the Mid-South derived from a Bayesian statistical model (Bissett and Miller 2017).

similar to Clovis but defined by an overall triangular form that is widest at the base. These points often have indented bases and more extensive fluting than Clovis bifaces (Goodyear 2006; Justice 1995:22). Albert C. Goodyear argued that this is indirect evidence for "instrument assisted fluting" (Daniel and Goodyear 2006; Goodyear 2006). Moreover, he contends that these points are likely post-Clovis in age based on similarities to sites in northeastern North America, such as Vail, Bull Brook, and Debert. However, almost all of the points come from surface contexts or private collections, and there is no site that firmly establishes their antiquity in the Mid-South. For the purposes of this study, Redstone and Gainey bifaces are included in the same temporal category as Clovis.

The second group includes the Cumberland and Barnes types, which are narrower and have large basal concavities. Cumberland bifaces are known for their slightly waisted appearance, whereas Barnes points are

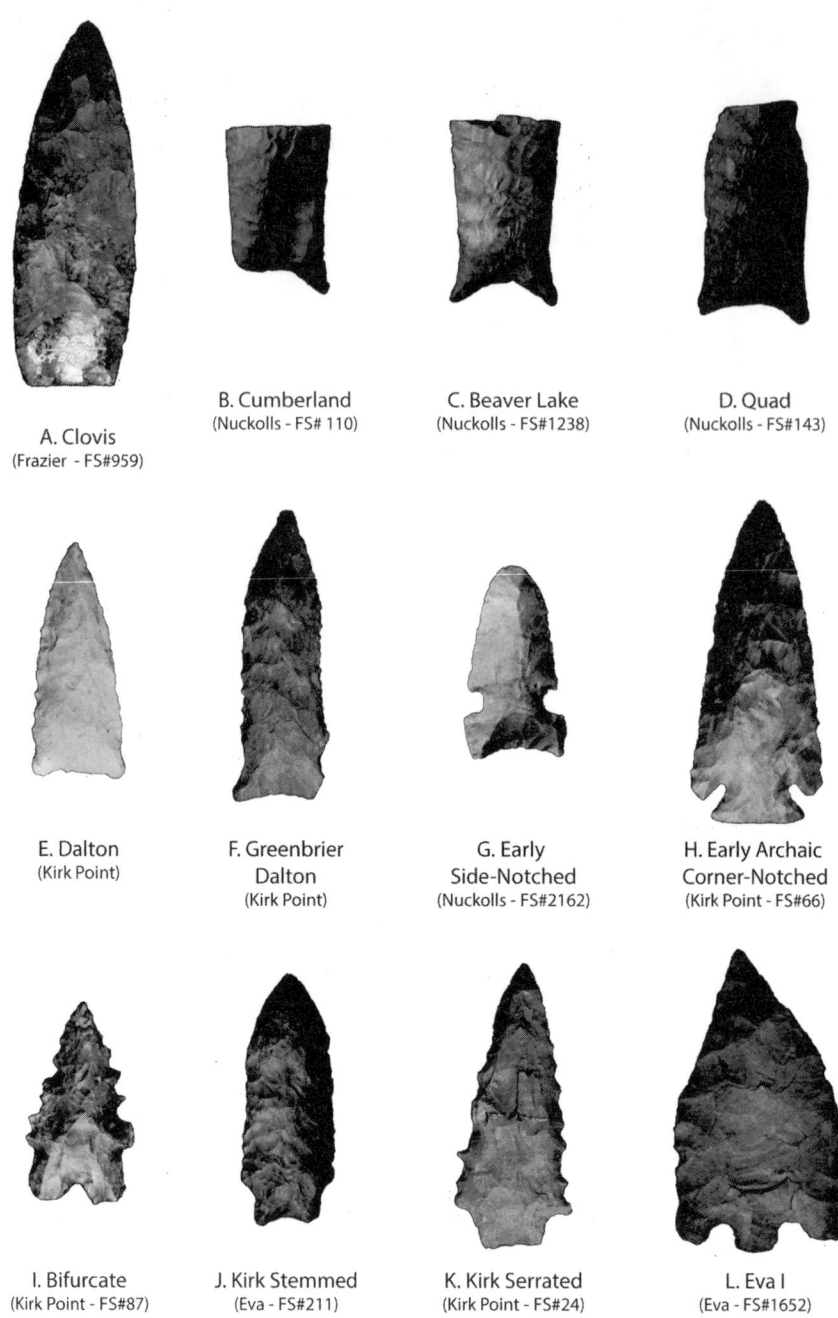

A. Clovis
(Frazier - FS#959)

B. Cumberland
(Nuckolls - FS# 110)

C. Beaver Lake
(Nuckolls - FS#1238)

D. Quad
(Nuckolls - FS#143)

E. Dalton
(Kirk Point)

F. Greenbrier
Dalton
(Kirk Point)

G. Early
Side-Notched
(Nuckolls - FS#2162)

H. Early Archaic
Corner-Notched
(Kirk Point - FS#66)

I. Bifurcate
(Kirk Point - FS#87)

J. Kirk Stemmed
(Eva - FS#211)

K. Kirk Serrated
(Kirk Point - FS#24)

L. Eva I
(Eva - FS#1652)

CM

FIGURE 2.9. Selected Paleoindian, Early Archaic, and Middle Archaic projectile point types from Benton and Humphreys counties, Tennessee.

more parallel sided. Additionally, the bases of Cumberland points are slightly concave and often have faint ears (Lewis 1954a, 1954b; Justice 1995:25–27) (Figure 2.9b). This group of points bears some resemblance to the comparatively well-dated Folsom points of the Great Plains. Here again, both Cumberland and Barnes points are found almost exclusively in surface contexts and private collections. Optically stimulated luminescence (OSL) dates are available at the Phil Stratton site in Kentucky (Gramly 2009), but the standard deviations are comparatively large, which makes it difficult to accurately assess the temporal relationship of this group of points with Clovis and other types. However, a heavily reworked Cumberland point was found at the deepest level of Dust Cave, stratigraphically within and below deposits containing Quad, Beaver Lake, and Dalton projectile points (Sherwood et al. 2004). Consequently, there is some basis to the claim that Cumberland predates these types and likely occurred between 10,800 and 10,500 ^{14}C (12,677–12,483 cal BP).

Goodyear (1999) and Anderson (2001) place both Quad and Beaver Lake points as the next types in the culture-historical sequence for the Mid-South. Beaver Lake points are slightly waisted lanceolates with faint ears, slightly concave bases, and moderate basal thinning (Cambron and Hulse 1975; Justice 1995:35–36) (Figure 2.9c). Quad points have distinct ears, a concave base, and pronounced basal thinning (Cambron and Hulse 1975; Justice 1995:35–36) (Figure 2.9d). While dates have been reported for these point types at Rodgers Shelter (Crane and Griffin 1972: 159) and Olive Branch (Gramly and Funk 1991), the key site for these points is Dust Cave in northern Alabama. Here, several distinct deposits were identified, and a relatively large number of AMS dates obtained (Sherwood et al. 2004). At this site, Quad and Beaver Lake points were found to co-occur, which is also reflected in their near identical spatial distribution across the region (Anderson et al. 2010). As a result, for this study, I will consider them as a single group. However, at Dust Cave, Sherwood and colleagues (2004) combined the dates from the Beaver Lake, Quad, and Dalton component. Bissett and Miller (2017) place the one-sigma range for the Quad and Beaver Lake types between 12,402 and 12,170 cal BP.

Goodyear (1982) argued that, in eastern North America, Dalton projectile points straddle the Pleistocene/Holocene boundary. Goodyear based his temporal assessment on the dates from Rodgers Shelter in Missouri and the Stanfield-Worley Bluff Shelter in northwestern Alabama. This sequence has been subsequently supported by excavations at Dust Cave, where Quad and Beaver Lake components were found stratigraphically below the Dalton components (Sherwood et al. 2004). Dalton bifaces typically begin life as lanceolates with concave bases and serrated edges. Often the basal margins of unreduced specimens are parallel to slightly incurvate, while the blade portion is initially excurvate (Justice 1995:40–42) (Figure 2.9e). Several studies have shown that the blade margins transition from excurvate to incurvate through repeated resharpening (Goodyear 1974; Shott and Ballenger 2007).

Another Late Paleoindian point type, Greenbrier, is described as a "lanceolate-bladed expanding stem form that shares characteristics with Daltons and later notched point types" (Justice 1995:42). This includes resharpening patterns characteristic of Dalton bifaces, although Greenbrier points also feature shallow side notches (Figure 2.9f). However, there are currently no published dates associated with these projectile point types, and based purely on morphology, this type likely postdates Dalton but precedes the side-notched and corner-notched types prevalent during the Early Archaic. For this study, I place both Dalton and Greenbrier Dalton in the same temporal category. Bissett and Miller (2017) place the one-sigma range for the Dalton and Greenbrier types as being between 11,771 and 11,441 cal BP.

After Dalton, a horizon of side-notched points occurred across the southeastern United States (Anderson 2001; Goodyear 1999) and into the southern plains with the notched varieties of the San Patrice type (Jennings 2008). Across the southeastern United States, the names of the side-notched varieties vary depending on who identified them first. In Florida and the Southern Coastal Plain, they are known as Taylor Side-Notched after Ripley P. Bullen's (1975) description of point types in Florida. However, in the Mid-South, they are known as Big Sandy points based on Cambron and Hulse's (1969) projectile point guidebook (Figure 2.9g). Subsequent summaries collapse these subregional types into a general Early Archaic Side-Notched horizon. In the Mid-South, these were first observed in early contexts at Stanfield-Worley, and again at Dust Cave (Randall 2002; Sherwood et al. 2004), as well as a handful of other sites. However, adding to the confusion is that side notching reappears in the Late Archaic, including at the regionally famous Big Sandy site in West Tennessee (Bissett 2014; Osborne 1942).

Consequently, the mere presence of side notching does not automatically make a point Early Holocene in age. Like the Dalton type, there are only a few sites with Early Archaic Side-Notched points with associated radiocarbon dates (n = 3). Bissett and Miller (2017) place the one-sigma range for the Early Archaic Side-Notched type as being between 11,384 and 10,588 cal BP.

Across eastern North America, a horizon of corner-notched bifaces appears during the Early Holocene (Anderson 2001). These are generally described as exhibiting a "large triangular blade with a straight or slightly rounded base and bifacially serrated blade edges" (Coe 1964:69–70; Justice 1995:71) (Figure 2.9h). These were found in stratigraphic context first at the Hardaway and Doershock sites in North Carolina (Coe 1964). This type was later dated at notable sites such as Icehouse Bottom and Rose Island in Tennessee (Chapman 1975) and St. Albans in West Virginia (Broyles 1966). These early dates have also been replicated at a series of other sites across eastern North America. In some instances, this type has been further split into a series of subtypes such as Kirk, Palmer, St. Charles, Lost Lake, Pine Tree, Cypress Creek, and Decatur based on variation in size, shape, notching, basal grinding, and flaking patterns (Justice 1995:71). However, for this study, these subtypes are lumped together into an Early Archaic Corner-Notched category. Bissett and Miller (2017) place the one-sigma range for the Early Archaic Corner-Notched type as being between 10,461 and 9,827 cal BP.

At St. Albans (Broyles 1966) and Icehouse Bottom (Chapman 1976), projectile points with bifurcated bases appear in deposits stratigraphically above those with corner-notched bifaces (Broyles 1966:23; Justice 1995:86–95; Lewis and Kneberg 1959) (Figure 2.9i). They have been classified into several subtypes, including MacCorkle, St. Albans, and LeCroy and have been dated elsewhere, where they consistently postdate components with Early Archaic Corner-Notched bifaces. For this study, I lump these point types together into a single Early Archaic Bifurcate category.

As opposed to the corner-notched base points that are widely distributed across eastern North America, bifurcated base projectile points are more spatially restricted, their range roughly corresponds to the Appalachian Mountains and mid-continent, and they are absent from the Atlantic and Gulf Coastal Plain (Justice 1995:89, 92). In other locations, stemmed-base varieties appear, including Kirk Stemmed, Kirk Serrated, Stanly Stemmed, and Kanawha Stemmed (Justice 1995:82–85, 95–99). At Hardaway and Doershock, these types are found stratigraphically above Kirk Corner Notched points (Coe 1964:69–70). However, at Icehouse Bottom and St. Albans, they are above the components with bifurcate-based points; and based on a series of sites across eastern North America, they consistently postdate both corner-notched and bifurcate-based points. Bissett and Miller (2017) place the one-sigma range for the Early Archaic Bifurcate type as being between 9,717 and 9,197 cal BP.

The next point types in the sequence are the stemmed varieties that appear at the end of the Early Archaic. The two most prominent types are Kirk Stemmed and Kirk Serrated, which were both identified at the Hardaway site (Coe 1964:70) (Figures 2.9j and 2.9k). Kirk Stemmed bifaces are described as having a long, narrow blade with a hafting element that is stemmed—likely by forming a broad notch opening that leaves a slightly expanding stem (Justice 1995:82). Kirk Serrated points are similar but with a more pronounced serration on the blade element (Coe 1964:70; Justice 1995:82).

Two other stemmed projectile point types that date to the terminal Early Archaic and early Middle Archaic are Kanawha Stemmed and Stanly Stemmed. Kanawha Stemmed points were identified at St. Albans (Broyles 1966:27), and they were found stratigraphically above deposits containing bifurcate types at Icehouse Bottom (Chapman 1975:125–126). The projectile points are described as having triangular blades and stemmed bases that sometimes have a short, shallow bifurcated base (Broyles 1966:27; Justice 1995:95). These points bear some resemblance to the Stanly Stemmed type, which, like the Kirk Stemmed and Kirk Serrated types, was also identified at the Hardaway site in North Carolina (Coe 1964:35–36). For this type, they are described as having "broad, triangular blades and narrow squared stems with shallow basal notched or bifurcation" (Coe 1964:35–36; Justice 1995:97). Based on their stratigraphic positions at Icehouse Bottom, St. Albans, and Hardaway, Stanly Stemmed points are considered to straddle the Early Archaic/Middle

Archaic temporal boundary. The radiocarbon dates for these points are limited to a handful of sites and overlap considerably in their morphology. As a result, for this study, I will consider them all as a single Early Archaic Stemmed group. Bissett and Miller (2017) place the one-sigma range for the Early Archaic Stemmed type as being between 9,408 and 8,308 cal BP.

The next group of projectile points consists of the Eva I, Eva II, and Morrow Mountain types. Lewis and Lewis (1961:40) identified both of the Eva I and Eva II types at the Eva site. Eva I points are "large, basally notched forms produced from a percussion flaked, triangaloid preform with a relatively straight basal edge" (Justice 1995:100) (Figure 2.9l). Eva II points are smaller and have straight or excurvate blades. However, the basal notching is the same as Eva I points (Coe 1964:37–43; Justice 1995:103) (Figure 2.10a).

While Eva I and Eva II points are found almost exclusively in the Mid-South, Morrow Mountain points have a much broader range and were first identified at the Hardaway site (Coe 1964:37–43). These points are described as a "small point with a broad triangular blade and a short, pointed contracting stem" (Justice 1995:104) (Figure 2.10b). At the Eva site, Eva I points are primarily found in Stratum IV, stratrigraphically below strata containing Eva II bifaces, and they also predate other sites containing Eva II and Morrow Mountain points in eastern North America (Bissett 2014). Eva II points were found associated with Morrow Mountain type points at the Eva site, and elsewhere across the eastern United States, the associated radiocarbon dates for the two points overlap, which has led most analysts to combine them into a single type (Sassaman 2001:230–231). Similarly, for this study, I treat Eva I and Eva II/Morrow Mountain types as two separate groups in Chapter 4 and combine them in Chapter 5. Bissett and Miller (2017) place the one-sigma range for the Eva I type as being between 8,687 and 8,313 cal BP and the Eva II/Morrow Mountain type as being between 7,570 and 6,914 cal BP.

While Morrow Mountain type bifaces are found across most of the southeastern United States (Sassaman 2001:230), in the Mid-South, the four point types that immediately postdate Morrow Mountain are more geographically restricted. These include the Sykes type, which are "broad, short-stemmed forms produced from the corners of trianguloid

A. Eva II
(Eva - FS#618)

B. Morrow Mountain
(Eva - FS#1397)

C. Sykes
(Eva - FS#1707(14))

D. White Springs
(Eva - FS#79)

F. Benton (Small)
(Eva - FS#1350)

G. Late Archaic
Side-Notched
(Eva - FS#861)

H. Ledbetter
(Ledbetter - FS#182)

E. Benton (Large)
(Eva -FS#170)

CM

I. Little Bear Creek
(Eva - FS#157)

J. Wade
(Eva - FS#275)

FIGURE 2.10. Selected Middle and Late Archaic projectile point types from Benton and Humphreys counties, Tennessee.

preforms" (Justice 1995:108) (Figure 2.10c). These points were first identified at the Eva site in the lower Tennessee River valley (Lewis and Lewis
1961:40–41), along with the White Springs type, which have "all of the
essential attributes of Sykes, but White Spring is more refined" (Justice
1995:108) (Figure 2.10d). Both types have been found in Middle Archaic
contexts at the Eva site (Lewis and Lewis 1961:40–41) and the Stanfield-
Worley Rockshelter (DeJarnette et al. 1962:70). Compared to the Eva and
Morrow Mountain types, there are comparatively fewer sites with ^{14}C
dates for the Sykes and White Springs types, yet the central tendencies
for the available dates clearly postdate the Eva/Morrow Mountain temporal distribution. Bissett and Miller (2017) place the one-sigma range
for the Sykes/White Springs type as being between 6,789 and 6,445 cal BP.

Two other point types appear to postdate the Eva and Morrow Mountain types and appear to be coeval with the Sykes and White Springs
types. The Benton type was originally identified at the Eva site (Lewis
and Lewis 1961:34) and described generally as large bifaces that are "easily
recognized by the presence of oblique parallel flaking that often occurs
on the blade" (Justice 1995:111) (Figures 2.10e and 2.10f). These projectile
points are stemmed, sometimes with large side notches; in some cases,
there is corner notching, like at the Ensworth site, which is located along
the Harpeth River in the Nashville Basin (Deter-Wolf 2004).

Charles H. McNutt (2008) compiled a database of ^{14}C dates to argue
that Benton projectile points postdate the Eva/Morrow Mountain points
and should be considered to mark the end of the Middle Archaic period.
Moreover, this type also appears frequently as caches in burial contexts
and may have been traded as part of an exchange network in the Mid-
South (Meeks 2000; Peacock 1988). While it is clear that this type postdates the Eva/Morrow Mountain type, at Eva they appear in the same
stratum (Stratum II—Big Sandy Phase) as the White Springs type. Also
occurring in this stratum is a smaller side-notched variety that Lewis and
Kneberg (1959:164) designated as Big Sandy points, although this has led
to some confusion, as similar side-notched forms also occur in Early Archaic contexts elsewhere in the southeastern United States (Justice 1995:
60–61; Figure 2.10g). However, at a handful of sites, side-notched forms
have been found to postdate the Eva/Morrow Mountain type and are coeval with the Sykes, White Springs, and Benton types. Bissett and Miller

(2017) place the one-sigma range for the Late Archaic side-notched type as being between 6,899 and 6,365 cal BP and the Benton type as being between 6,343 and 5,979 cal BP. For this study, these types and the Sykes/White Springs type are combined into a single Middle Archaic Stemmed category.

Following the Middle Archaic Stemmed group is another set of stemmed projectile points, the most notable of which is the Ledbetter type (Figure 2.10h). Named after the Ledbetter site in the lower Tennessee River valley, it was also found in Late Archaic contexts at the Eva site (Lewis and Lewis 1961) and at Russell Cave in northern Alabama (Ingmanson and Griffin 1974:45). These points are described as "a contracting stem form with an asymmetrical blade" and usually have barbed shoulders (Justice 1995:65). This type is also known as the Pickwick type (Cambron and Hulse 1969:75) and is similar to the Etley type, which is found in Late Archaic contexts in the mid-continent (Justice 1995: 146–148). An additional stemmed type, Little Bear Creek, also overlaps temporally with the Ledbetter type and is described as "medium to large with slightly to excurvate blade edges and has a long stem" and "the haft element varies from straight to contracting stem with grounding on the lateral margins of the stem" (Justice 1995:96) (Figure 2.10i). While Noel D. Justice (1995:196) places this type in the Early Woodland period, most of the available dates for this type are coeval with the Ledbetter and Pickwick types and place it firmly in the Late Archaic. For this study, these points will be combined into a single Late Archaic Stemmed category. Bissett and Miller (2017) place the one-sigma range for the Late Archaic Stemmed type as being between 4,740 and 3,542 cal BP.

The Wade projectile point type appears next in the chronological sequence and is "basically straight-stemmed forms with wide barbs" (Cambron and Hulse 1969:110; Justice 1995:180) (Figure 2.10j). These points appear to straddle the Late Archaic/Early Woodland boundary (e.g., Faulkner and McCollough 1973:149) and have been found in preceramic contexts in shell middens as well as associated with Early Woodland pottery (Justice 1995:180). These points are morphologically similar to the Delhi type, which are "straight stemmed forms with barbed shoulders" (Ford and Webb 1956; Justice 1995:79). This type has been found at the

Poverty Point site (Ford and Webb 1956:117) and, like the Wade type, is found in Late Archaic and Early Woodland contexts in the Mississippi River valley and Mid-South (Justice 1995:179). Following Justice (1995: 179–180), I combine these types into a single Late Archaic Barbed category. Bissett and Miller (2017) place the one-sigma range for the Late Archaic Stemmed type as being between 3,156 and 2,766 cal BP.

Conclusion: Correlating Climate, Culture, and Bifaces in the Mid-South

Based on the information compiled here, it is possible to correlate major environmental and cultural trends with changes in the morphology of temporally diagnostic bifaces in the Mid-South during the Paleoindian and Archaic periods. Clovis bifaces appeared during the Bølling-Allerød period and disappeared along with a wide array of species (e.g., the Rancholabrean extinction) at roughly the onset of the Younger Dryas (Meltzer 2009:255–265). During the Younger Dryas, a series of more geographically restricted types appear (Anderson et al. 2010). In the Mid-South, these include Redstone/Gainey, Cumberland/Barnes, and Quad/Beaver Lake; the Dalton type appears at the end of the Younger Dryas.

During the Early Holocene, the climate became warmer and deciduous forests continued to expand northward (Delcourt and Delcourt 1985, Viau et al. 2006). Over this period, people produced a succession of point types, including Greenbrier, Early Archaic Side-Notched, Early Archaic Corner-Notched, Bifurcates, and Early Archaic Stemmed. Based on the distribution of archaeological sites in the Southeast, Early Archaic groups had clearly expanded across most of the Southeast (Anderson 1996), including the Appalachian Summit (Kimball 1996).

During the mid-Holocene, the climate became warmer and drier, which is reflected in the expansion of the prairie grasslands into the Midwest, cedar glades in the Nashville Basin, and the pine forests of the Atlantic and Gulf Coastal Plain (Delcourt and Delcourt 1985). It is during this time that Eva/Morrow Mountain and Middle Archaic Stemmed types were made.

At the end of the mid-Holocene, the climate cooled slightly, which is reflected locally by the retreat of the prairie grasslands in the Midwest and the shrinking of the cedar glades in the Nashville Basin. In the Mid-South, the Late Archaic Stemmed and Terminal Archaic Barbed types overlap with all of these trends.

The following chapters will explore ways in which formal models from behavioral ecology can be used to track variation in bifaces, re-sharpening, and the distribution of archaeological sites to make inferences regarding hunting returns and demographic trends in the lower Tennessee and Duck River valleys.

Notes

1. Unless noted otherwise, all dates are calibrated using OxCal 4.2 and the IntCal09 curve (Bronk Ramsey and Lee 2013) and presented as calendar years before present (cal BP).

CHAPTER 3

From Projectile Points
to Prey Size

Introduction

If there is one stone tool that is present in most prehistoric archaeological assemblages in eastern North America, it would be bifaces. Bifaces are "lithic tools that have been extensively modified by chipping and have two sides or faces that meet to form a single edge that circumscribes the entire specimen" (Andrefsky 2006:744). They are one of the most intensively studied artifact classes in eastern North America, especially for the Paleoindian and Archaic periods. This is largely due to the fact that their sizes and shapes vary predictably with time, making them fairly reliable temporal markers in a region where there is poor preservation of organic material for ^{14}C dating (Anderson and Sassaman 2012:5; Dunnell 1990:13). Moreover, as demonstrated in the previous chapter, there is a loose correlation between changes in climate, culture, and biface morphology (Anderson 2001).

Bifaces have also played a prominent role in studies of technological organization in North America. Binford's dichotomy between forager and collector strategies, as well as other publications based on his ethnoarchaeological fieldwork, provided a theoretical foundation for interpreting lithic assemblages (e.g., Binford 1979, 1980; Kelly 1992). Some more notable early attempts to apply this framework to archaeological case studies in the southeastern United States include Anderson and Hanson's (1988) Early Archaic band-macroband settlement model for the Atlantic Coastal Plain, Larry R. Kimball's (1981, 1996) examination of Early Archaic site use in the Little Tennessee River valley, and Daniel S. Amick's (1987) raw material survey of the Duck River valley. In each of these examples, bifaces were the primary artifacts analyzed, and they have continued to play a central role in the study of lithic technological organization in the southeastern United States (Carr and Bradbury 2000; Carr et al. 2012).

The application of narrative models derived from Binford's work also began to proliferate in the Paleoindian literature due to a handful of influential studies. These include Goodyear's (1979, 1989) cryptocrystalline hypothesis, in which he argued that Paleoindian groups targeted high-quality raw materials that were often moved great distances from their source. Because of the quality of the material, artifacts could be resharpened, reused, and recycled repeatedly, which would have provided a high degree of flexibility in the toolkits of highly mobile hunter-gatherer groups reliant on curated tools. Similarly, Kelly (1988) argued that bifaces can play three roles in the organization of technology—cores, bifacial tools, and preforms for projectile points. In the same year, Kelly and Todd (1988) incorporated part of this hypothesis into their high-technology forager model for the colonization of North America, where they argued that a reliance on bifacial technology was part of a versatile mobility strategy that can be adapted to novel environments.

With the crystallization of the technological organization theoretical perspective (e.g., Nelson 1991), analysts shifted their attention to identifying the various factors that could influence decisions regarding the manufacture, maintenance, discard, and design of stone tools, and in particular bifaces and endscrapers (Andrefsky 2009:70–72, and references therein). These include access to raw material (Andrefsky 1994; Bamforth 1986; Ingbar 1994; Kuhn 1995, 2004), the role of time and risk (Torrence 1983), maintainability and reliability (Bleed 1986), and other factors (e.g., Andrefsky 2009:71).

In recent years, there has been an attempt to move away from narrative explanations to formal models that are on more solid quantitative footing. Examples would include Steven L. Kuhn's (1994) formal model for lithic tool transport, which predicts that highly mobile hunter-gatherers transport finished or near-finished tools rather than cores. Similarly, P. Jeffrey Brantingham (2006) generated a neutral model of tool discard based on a Lévy random walk as a baseline from which to interpret raw material variation in assemblages in southern France. These two archaeologists constructed null models based on an economically rational baseline and then compared archaeological data to these results.

In other instances, analysts have derived null hypotheses directly from the formal models of behavioral ecology. Beck and colleagues (2002) used a central place provisioning model from Metcalfe and Barlow (1992) to frame hypotheses for lithic provisioning in the Great Basin. Todd A. Surovell (2009) created a set of economic models as a means to analyze differences in occupation intensity at a sample of Folsom sites on the Great Plains and Rocky Mountains. Kuhn and Miller (2015) derived expectations for the discard of artifacts based on the marginal value theorem (MVT) (e.g., Charnov 1976) by treating artifacts as patches of potential utility (see also Burger et al. 2005; Shott 2015; Sih 1980).

These attempts to adapt models from behavioral ecology to lithic technology provide important case studies in how archaeologists have used the stone tool record to track changes in mobility and subsistence over time. In Chapter 1, I discussed how Stiner (2001) used the diet breadth model to construct a null model that allowed her to argue that population pressure was a causal factor in the broad spectrum revolution that preceded the origins of agriculture in the eastern Mediterranean. While faunal preservation is much more problematic in eastern North America, this is offset by an abundant lithic record. Consequently, a formal model is required that allows analysts to translate variation in the size, shape, and condition of bifaces that were likely used as projectile points into inferences about the tasks for which those projectile points were likely used: hunting.

Projectile Points as Patches

In a review of the study of lithic technology, William Andrefsky, Jr. (2009) divided the literature along basically two axes of variation that archaeologists can explore: 1) variation in the raw material used to make stone tools and 2) the use-lives of the tools that are fashioned out of the raw material. For the former, the studies focused on factors that included access to raw material, the quality of the material, and the cost of producing a tool as measured in time or energy. For the latter, an artifact's use-life would be contingent on the variety of tasks for which a tool could be used, and those tasks would, in turn, influence the probability of catastrophic failure (i.e., breakage that would render a tool unable to perform a desired task) and the attrition rate (i.e., mass lost due to use and maintenance). These factors would then be weighed against the actual energy gained from using the tool for a task versus an alternative tool or no tool at all.

Framed in this way, an individual's estimation about the availability of raw material and the cost of producing a new artifact is juxtaposed against the expected gains, measured in caloric return, from a new artifact. As a stone tool producer, an individual must consider *energy lost* due to extraction and transport costs when acquiring suitable raw material and the costs associated with producing and using a new tool. Conversely, as a stone tool consumer, an individual must also consider the *energy gained* by using a particular tool at a given time and how many times a tool can be used before being maintained or replaced. More simply, an individual must balance the amount of energy lost to production costs to the energy gained from choosing to use, or consume, a tool.

Kuhn and Miller (2015) applied many of the same concepts in a formal model for the prediction, maintenance, and discard of artifacts that was derived from the marginal value theorem. They argued that artifacts can be considered "patches of utility" (Figure 3.1), where the criterion E is the expected yields for the replacement artifacts minus the replacement costs, which is then divided by the time used or number of uses. The optimal point of artifact abandonment occurs when immediate returns drop below expected yields for replacement artifacts ($T1$, $T2$).

As a modern analog, imagine a hunter with a rifle, ammunition, and a permit to hunt deer. For the sake of the analogy, let's also assume the hunter's goal is subsistence, not mounting the animal on a wall for prestige. In this model, E is essentially the energy gained from using a given tool after balancing the cost of factors, like the cost of acquiring the gun and ammunition and travel time. How the hunter perceives these costs is going to impact behavior when he or she actually comes across a deer.

The rounds of ammunition in the gun can be considered "patches of utility," and how quickly a hunter moves from patch to patch (or shot to shot) can be modeled using the MVT, assuming the following constraints derived from Kuhn and Miller (2015:178): 1) Hunters effectively use a single round at one time, 2) the utility of artifacts declines with

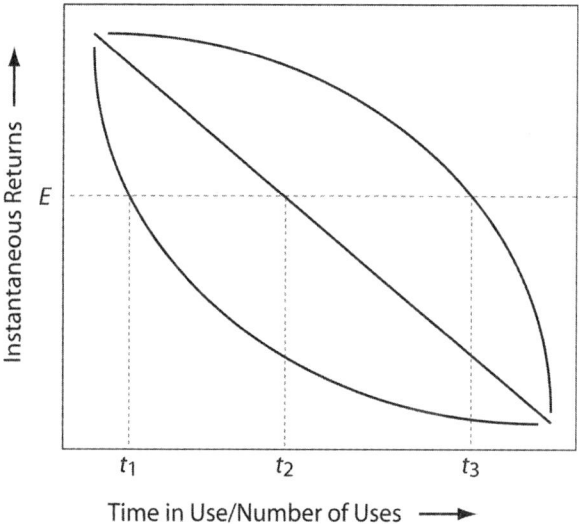

FIGURE 3.1. The marginal value theorem applied to stone tool production, use, and discard (Kuhn and Miller 2015). *E* is the criterion
for artifact abandonment, which is equal to the expected yields for
replacement artifacts minus amortized replacement costs, divided by
the number of use events. The optimal points of artifact abandonment
(*t1, t2, t3*) occur when immediate returns drop below expected yields
for replacement artifacts. The upper, middle, and lower curves reflect
different attrition rates for the use of artifacts over time.

successive uses, 3) people monitor artifact effectiveness or the condition
of the ammunition, and 4) people are aware of the costs of replacing
ammunition. For the sake of our caricature, let's assume that when a
hunter encounters a deer, the best shot is the first shot, and the odds of a
successful shot decline as the deer is running away.

Consequently, the hunter has to balance how likely they are to hit a
deer with that first shot, how likely they are to get off a second successful
shot if they miss the first shot, and, just as importantly, how much ammunition they have and how much it costs to replace it. On one hand, if
the hunter fires at a herd of deer and ammunition is cheap, it's likely that
a hunter would not be concerned about firing multiple times—replacement costs are low and the odds of getting off a second successful shot
are pretty good. Conversely, if ammunition is expensive and the hunter
only encounters one deer, it's more likely the hunter would conserve

rounds in this context. Or, to put it another way, a hunter would exercise more caution to make sure that first shot is taken at precisely the right time that maximizes the probability of success.

To parse this out further, Kuhn and Miller (2015:181–182) argue that modeling tools as patches of utility leads to two obvious conclusions and one that is not so readily apparent. First, when the costs of a tool increase, then that should increase the number of times an artifact is used. Second, declining utility should increase the speed of abandonment. Finally, although it may appear counterintuitive at first, if the returns from using a tool increase, that should lead to an increase in the frequency that the tool will be replaced, as well as the inverse that if yields decline, the artifact should be retained for a longer period of time.

Returning to the anecdote of the rifle-wielding hunter, if ammunition is expensive, it should be expected that the hunter should use it, or else why buy it and carry it around? On the other hand, if the hunter purchased a large allotment of rounds and then realized that there are very few deer, with limited opportunities for multiple shots upon encountering a deer, the hunter might choose to abandon those superfluous rounds rather than carry them around.

Conversely, if the hunter discovered that deer were encountered frequently in large groups, it might be worthwhile to carry more ammunition and take more shots to ensure a successful encounter, which would lead to the hunter needing to restock their ammunition supply more frequently. The inverse expectation would be that if deer were encountered infrequently, the hunter would probably hang onto those initial rounds for much longer. Consequently, if the hunter left behind expended cartridge rounds, an archaeologist could begin to make inferences about the behavior of the hunter if the cost of the ammunition and/or the likelihood the hunter would encounter a deer could be estimated.

Now imagine that the anecdotal hunter is no longer carrying a rifle but instead an atlatl (or spear-thrower). Rather than firing bullets, they are instead throwing spears, which may or may not be tipped with a stone projectile point. Instead of trying to make inferences about the hunter based on discarded cartridges, archaeologists would have to instead make inferences based on either the stone projectile points (which could be recycled, reworked, and also used as knives—unlike bullets) or the tools that were used to produce spear shafts (which could be used to produce things other than spear shafts). This presents two related issues for archaeologists attempting to translate Kuhn and Miller's (2015) artifacts-as-patches model to the archaeological record:

1. How can analysts untangle lithic raw material quality and availability and artifact life histories as two axes of variation in archaeological assemblages?
2. What are reasonable proxies for measuring energy expenditure with archaeological data?

One strategy for controlling raw material constraints is to analyze assemblages in areas with ubiquitous, high-quality raw material (e.g., Andrefsky 1994:30) or stable, permanent locations on the landscape that were likely used repeatedly over time, like caves or rockshelters. Alternatively, an analyst could control for raw material constraints by examining material from known sources and regressing the distance from site to source against other variables of interest (e.g., Eren and Andrews 2013).

Surovell (2009) used a 20 km buffer around his sites, which is the average daily maximum foraging radius for ethnographic hunter-gatherers. Materials from sources beyond 20 km are considered extra-local and are therefore more costly to acquire. This provides a proverbial line in the sand for comparing the frequency of raw material types in different locations. These various analytical strategies provide archaeologists a means to either minimize the effects of or explore the costs associated with producing new tools.

With stone tools, archaeologists analyze the amount of lithic raw material that is consumed, or more specifically, how many times a tool has been used before breakage or discard. One artifact class that has received a considerable amount of attention in this regard has been scrapers or retouched flakes, where several methods have been derived for determining the size of the original flake prior to retouching relative to the mass of the artifact upon discard (Clarkson 2002; Eren and Prendergast 2008; Hiscock and Clarkson 2005; Kuhn 1990). These researchers are trying to determine the "expended utility" for each artifact (e.g., Shott 1996:267). Similar attempts have been made to develop ways to measure the effects of breakage and resharpening on bifaces, and in particular projectile points (Andrefsky 2006; Goodyear 1974, 1979; Hoffman 1985; Miller and Smallwood 2012; Shott and Ballenger 2007; Wilson and Andrefsky 2008). For example, Amick (1996) used the ratio of broken to complete bifaces in his sample of Folsom assemblages, which can be considered an index of catastrophic breakage.

For measuring resharpening, most studies examine the relationship between the parts of the biface that have been hafted relative to the total length. This is because hafted portions of bifaces are less likely to be reworked (e.g., Keeley 1982), which means that the length will be preferentially removed during resharpening. Kuhn and Miller (2015) simply used the maximum length divided by the maximum width as a proxy for resharpening. This ratio is effective because it is an easily measured proxy for how much of the length of the biface has been reduced by resharpening or repair.

Recently, several experimental studies have tested the assumptions that underpin these archaeological proxies for projectile point resharpening and breakage. Cheshier and Kelly (2006) fired obsidian-tipped arrows at a deer carcass and found that roughly half broke upon first use, and overall, the shorter and thicker points were significantly more durable. They argued that their results are consistent with the high failure rates in comparable studies by Titmus and Woods (1986) and Odell and Cowan (1986). Moreover, Waguespack and colleagues (2009) fired arrows tipped with stone and wood at ballistics gel covered in hide, and they found that stone-tipped projectile points only penetrated 10 percent deeper than wood-tipped points. They argued that if this translated to greater blood loss, using a stone-tipped point could convey some economic benefit with larger prey, which is consistent with some ethnographic case studies where stone-tipped spears and arrows were used preferentially for larger game and close-range kills. However, they remain skeptical that this payoff would be enough to warrant the use of stone-tipped projectiles over their wooden-tipped counterparts.

David A. Hunzicker (2008) examined both the probability of catastrophic failure and resharpening on projectile points. In his experiment, he fired 25 Folsom point replicas attached to atlatl darts in five different hafting designs into bovine rib cages using a tripod-mounted crossbow.

He found that three-fourths of the firing attempts successfully penetrated the target, and after resharpening, the damaged points averaged
five attempts before being broken beyond repair. He also found that after
a total of 108 firing attempts, 35 missed the ribs and suffered no visible
damage. However, of the 73 attempts that made contact with ribs, all were
damaged, including 18 attempts where the points were irreparably damaged. Half of the breaks were snaps or crushes that allowed the points to
be resharpened, as opposed to longitudinal fractures. Finally, he argued
that the reduction in length was the best proxy for measuring point rejuvenation, as opposed to thickness, width, and a width-to-thickness
ratio, which did not appear to vary significantly. Hunzicker then plotted
length against the number of rejuvenation attempts to create an experimental attrition curve and compared these results to published Folsom
assemblages. He argued that the Folsom assemblages likely reflect greater
amounts of use and resharpening than previously recognized.

The attrition rates observed by Hunzicker were drastically different than those observed by Cheshier and Kelly (2006)—likely because
Hunzicker used projectile points made of chert rather than obsidian and
fired his points at the end of atlatl darts at a lower velocity than the bow
and arrow that Cheshier and Kelly used in their experiment. Iovita and
colleagues (2014) used replicas of Levallois points cast in glass and fired
them into ballistics gel and synthetic bone using an air gun. They found
that as kinetic energy increases the probability of breakage increases,
and the breaks are more likely to be longitudinal fractures rather than
transverse snaps. Consequently, it should not be surprising that firing
points made out of a more brittle raw material at a much higher velocity
into a target with more closely spaced ribs should experience higher rates
of catastrophic failure.

These factors may also have influenced the decisions of some groups
to invest in stone-tipped projectile points. In the study by Waguespack
and colleagues (2009), they used ballistics gel with no proxy for bone,
whereas Hunzicker showed in his study that every point that made
contact with bone was damaged. While Hunzicker's points were reju-
venated after damage, there is no comparable study to test the efficacy
and durability of wooden points against targets with bone. Moreover,

Waguespack and colleagues argue that there is relatively little cost associated with wood-tipped arrows or spears compared to their stone-tipped counterparts.

However, Richard Gould (1980), in his ethnographic accounts of the Ngatara in Australia, observed that individuals employed a wide array of tools to produce and maintain their wooden spears. Consequently, the rationale for choosing wood versus stone-tipped projectiles may involve a trade-off between carrying raw material around at the end of an arrow or spear or carrying raw material in the form of maintenance tools to produce and resharpen softer, wooden-tipped projectiles.

Based on Hunzicker's (2008) study, the assumptions that projectile point length is differentially affected by impact and resharpening and that these points can survive multiple rejuvenation attempts appear entirely reasonable. Moreover, the differences between Chesier and Kelly (2006) and Hunzicker (2008) show that there may also be reason to believe that variation in prey size (e.g., deer versus bovine targets) may have also differentially impacted the attrition rate and probability of catastrophic failure, which is consistent with the hypothesis by Kuhn and Miller (2015) that the increase in biface resharpening in the Southeast observed in Paleoindian points from Tennessee is due to a shift from hunting megafauna to deer and a wider array of smaller species (e.g., Morse 1973; Walker 2007). Furthermore, establishing a relationship between projectile point resharpening and prey size would be extremely important because it would provide an archaeologically tractable proxy for the amount of energy gained from the use of these particular tools. However, there are other confounding variables that also explain the differences between their results, including raw material selection, resharpening, point size, point shape, and firing velocity.

An alternative approach to demonstrating the relationship between projectile point resharpening and prey size (and, by extension, hunting returns) is to compare faunal remains and projectile points from archaeological contexts where access to raw material can be held constant. However, this is surprisingly difficult in eastern North America, where published data on sites containing deposits with preserved faunal remains are few and far between (Styles and Klippel 1996). Instead, I use two case studies—Puntutjarpa Shelter in Australia (Gould 1977) and

Gatecliff Shelter in Nevada (Thomas 1983)—to demonstrate that if *access to raw material is held constant*, then the resulting variation is likely due to *fluctuations in hunting returns*.

Puntutjarpa Rockshelter, Western Australia

Puntutjarpa Rockshelter in western Australia was excavated under the direction of Richard Gould from 1969 to 1970 and contains a record of occupation that spans the entirety of the Holocene epoch. The results of this meticulous excavation are provided in a full report that contains a highly detailed description of the lithic inventory recovered at the site as well as a subsequent analysis of the faunal remains (Archer 1977; Gould 1977, 1996). Surovell (2009) used the information from this excavation as an independent test of his site occupation index model and found that components with assemblages indicative of higher degrees of occupational intensity corresponded to spans of time with increased effective precipitation. This relationship may be driven by the observation that the largest animals—red kangaroo (*Megaleia rufa*), hill kangaroo (*Macropus robustus*), and emu (*Dromaius novaehollandiae*)—become more plentiful after seasons of heavy rains (Gould 1977). Moreover, Gould (1977:28–29, 169–170) argues that the availability of water is the single most important variable that influences residential mobility, group size, the likelihood of group hunts, and prey selection.

The Puntutjarpa Shelter dataset is also valuable because of the extensive ethnographic and ethnoarchaeological research among the Ngatara conducted by Gould (1980) prior to the excavations at Puntutjarpa Shelter. Consequently, Gould argues that it is possible, in a very general sense, to derive the function from the form of the artifacts from direct ethnographic observation. As a result, Puntutjarpa Shelter can be used as a case study for examining the relationship between hunting returns and projectile point resharpening. However, in this case, the Aboriginal Australians used nine-foot-long wooden spears that were hurled with spear-throwers, mostly from behind blinds, or other forms of stealth hunting (Gould 1977:26–28). Since the Aboriginal Australians who inhabited Puntutjarpa Shelter did not utilize stone projectile points, I instead use stone tool types that Gould argued are used almost exclusively for the production and maintenance of spear shafts (Table 3.1). These include

TABLE 3.1. Artifact classes and ethnographically observed functions from Gould (1977).

Type	Type Abbreviation	Function	Page #
Large cores	1a, 1c	Only the very hardest wood	92
Horsehoof core	1b	Chopper/scraper planes?	93
Large flake Scrapers/ spokeshaves	2a, 2b	Used exclusively for scraping hard woods to make spear shafts, whether for smoothing them or for shaping and sharpening	94
Adzes and adze slugs	3a–3d	Scraping hard woods	96–97
Micro-adzes and micro-adze slugs	3e–3h	Scraping hard woods	98
Small endscrapers	3i	Scraping hard woods	99
Backed blades	4	Multiple?	99–100
Handaxes	5	Early woodworking production	100
Ground seeder/ grinder	6	Grinding seeds	101
Grinding slabs	7	Grinding seeds	101
Hammerstone	8	Hammerstone	101
Retouched fragments and utilized flakes	9	Multiple?	101–102

one type he claims were used exclusively for sharpening and straightening spears (Types 2a and 2b).

With the faunal data, only presence and absence information by species was available (Archer 1977:159). I divide those into two groups: large species (the three species of kangaroo) and small species (everything else) (Table 3.2). I then standardize the number of small species by the bone weight and count for each level as a coarse proxy for species diversity (Table 3.3). I then regress these values against Surovell's (2009) occupation span index values and find that they correlate (Figure 3.2). In other words, as occupation intensity increases, faunal diversity decreases. I then compare the proxy for faunal diversity to the adzes that Gould described as woodworking tools. I divide the frequency of these by the number of utilized flakes and the total number of tools for each level in order to create a measure of the relative abundance of these tools compared to the overall lithic assemblage. Both proxies for the frequency of woodworking tools correlate with the proxy for small species diversity

(Table 3.4; Figures 3.3 and 3.4). I interpret this as indicating that there is a greater frequency of tools suggestive of woodworking activity occurring when there is a wider array of small species present. Finally, the frequency of these woodworking tools also correlates with the occupation span index for each component (Figure 3.5). Based on this analysis, it appears that the occupation span index, prey size, and the frequency of woodworking tools for each component are correlated.

These findings are consistent with Kuhn and Miller's (2015) artifacts-as-patches model. When return rates are lower (represented by a greater diversity of small species), there should be more maintenance of tools. In this case, the signature of more maintenance is the greater relative abundance of adzes, spokeshaves, and other tools that have been observed ethnographically as woodworking tools, including adzes hafted to spear-throwers that were used primarily for resharpening wooden spears (Gould 1980:14). Moreover, while the faunal data from Puntutjarpa Shelter is highly fragmented, the reporting of the lithic data is superb. When you compare the proposed woodworking tools to Surovell's (2009) occupation span index for each level, there is a significant negative correlation. The more time people spent in the shelter, the fewer woodworking tools they discarded relative to everything else.

This pattern conforms to many of Gould's (1980) ethnographic observations of contemporary Ngatara people, who inhabit the same region as Puntutjarpa Shelter. The only large animals available for hunting (kangaroo and emu) become noticeably more abundant and cluster in mulga woodlands with small flats of grass when precipitation increases (Gould 1980:192). These areas are, in effect, vegetation oases, or corridors. Puntutjarpa Shelter is located near one of these and has an extensive viewshed. Also, Gould (1980:66) argued that human group size increases (and aggregation events are more likely to occur) when hunting returns are higher, which also covaries with rainfall. Conversely, a decrease in rainfall would likely depress the density of kangaroo and emu, decreasing herd size for these animals, and lead to an increase in logistical foray distance for human groups. Moreover, hunters would have been more likely to widen their diet breadth to include smaller, harder-to-catch species.

The combined effect of decreasing rainfall and a widening diet breadth would have likely amplified the number of unsuccessful spearing

TABLE 3.2. Presence and absence of observed faunal taxa by excavation level from Puntutjarpa Shelter (Archer 1977; Gould 1977, 1996).

Taxa	Name	AX	AX-BX	BX	CX	CX-DX	DX	EX	EX-FX	FX	GX	GX-HX	HX	IZ	IZ-JZ	JX	KZ	KZ-LZ	MS	MS-NS	NS	NS-OS	OS	OS-PS	PS	PS-QS	QS	QS-RS	RS	RS-SS	SS	SS-TS	TS	TS-US	US	US-VS
Dasyurus cf geoffroii	Western native cat	X	X																																	
Perameles cf. P. eremiana	Desert bandicoot																										X									
Chaeropus ecaudatus	Pig-footed bandicoot			X																				X				X								
cf. *Isoodon auratus*	Golden bandicoot					X	X																													
Macrotis lagotis	Common rabbit-eared bandicoot	X					X				X																									
Trichosurus cf. T. Vulpecula	Brush-tailed possum	X			X	X	X	X	X	X	X	X	X			X			X	X	X		X	X	X			X	X							
Bettongia lesueur	Lesueur's rat kangaroo	X	X	X	X	X	X	X		X	X		X			X			X	X	X		X	X	X	X		X	X							
Lagorchestes hirsutus	Western hare-wallaby		X									X											X	X												
Onychogalea lunata	Crescent nail-tail wallaby			X	X							X	X				X					X		X			X	X				X				
Petrogale sp.	Rock-wallaby														X												X	X			X					
Macropus robustus	Large kangaroo						X	X								X	X					X						X								
Megaleia rufa	Large kangaroo				X	X	X	X	X			X	X	X		X	X		X	X		X	X	X	X	X		X								
Large macropodid	Large kangaroo		X	X	X	X	X	X	X	X	X	X	X	X	X	X	X		X	X		X	X	X	X	X	X	X	X	X	X	X	X	X	X	X
Muridae	Native mice	X	X	X	X		X	X													X															
Oryctolagus cuniculus	European rabbit	X	X	X	X			X	X																											
Large carnivore(s)	Dingo/Thylacine/Tasmanian devil				X	X					X		X	X									X					X						X	X	
Varanidae varanus	Monitor lizard			X			X													X																
Ophidia	Snakes																						X													
Aves	Birds	X																																		
Total		7	7	11	6	6	8	8	3	3	3	6	4	2	2	3	5	1	4	4	3	3	7	4	3	2	3	6	4	1	2	2	1	1	1	1

TABLE 3.3. Data for Puntutjarpa Shelter from Gould (1977, 1996) and Surovell (2009).

Stratum[1]	Area (m²)	Artifacts	Density	Occ. Span Index	Type 1 (Cores)	Type 2 & 3 (Woodworking)	Type 2 & 9 (Utilized Flakes)	Total Tools	Waste Flakes	Small Species	Large Species	Bone Frag.	Bone Wt. (g)
AX	15.88	770	48.5	27.15	2	9	8	17	753	5	2	1211	507.4
BX	15.26	1537	100.7	34.14	1	26	19	47	1590	9	2	776	322.1
CX	14.64	2622	179.1	29.14	4	49	36	92	2530	4	2	901	464.2
DX	14.02	3806	271.5	30.14	13	70	59	142	3664	6	2	853	507.2
EX	13.7	4535	331	27.69	15	82	66	168	4367	5	3	894	549.2
FX	13.69	3727	272.2	27.83	9	70	60	138	3589	2	1	1066	673.6
GX	13.68	4792	350.3	35.36	25	55	53	139	4653	2	1	1284	791.5
HX	13.67	3687	269.7	34.95	15	30	29	76	3611	2	2	893	540
JZ	10.03	3116	310.6	59.82	7	9	22	38	3078				
JZ	7.53	1335	177.4	50.96	5	17	10	30	1315				
KZ	5.02	1606	320.1	55.8	4	5	13	25	1591				
LZ	2.51	636	253.5			2	5	10	626				
M5	13.82	8146	589.4	79.08	14	19	44	85	8061	2	2	1505	883.5
N5	13.79	9845	714.1	85.46	13	16	34	81	9764	2	1	1954	988.5
O5	13.75	8768	637.5	74.22	22	17	33	77	8691	5	2	2163	1275.3
P5	13.72	6140	447.5	74.93	20	7	21	54	6086	1	2	1701	1031.5
Q5	13.65	4657	341.1	93.78	11	18	37	68	4589	2	1	811	489.2
R5	13.55	3790	279.6	57.32	13	10	21	45	3745	2	2	646	345.5
S5	13.46	1908	141.8	44.05	7	6	17	30	1878	1	1	310	136.6
T5	13.36	1411	105.6	30.64	8	8	25	42	1369	0	1	174	100.8
U5	10.63	456	42.9	17.63	6	5	12	22	434	0	1	23	10.2
V5	5.24	76	14.5	14.79		0	3	5	71	0	1	1	0.1
X5	2.73	18	6.6			0			18			1	0.2

[1] Strata JZ–LZ consists of large pieces of roof fall that the principal investigators deemed problematic and excluded from their analysis (Gould 1996). Zones T5–X5 had no small species present and had low values for bone weights compared to the rest of the strata, which is why they were excluded.

TABLE 3.4. Proxies for Puntutjarpa Shelter derived from Gould (1977, 1996) and Surovell (2009).

Stratum	Occupation Span Index	Small Species: Bone Weight	Small Species: 1k Fragments	Adzes: Utilized Flakes	Adzes: Total Tools
AX	27.15	9.854158	4.128819	1.125	0.529412
BX	34.14	27.94163	11.59794	1.368421	0.553191
CX	29.14	8.616975	4.439512	1.361111	0.532609
DX	30.14	11.82965	7.033998	1.186441	0.492958
EX	27.69	9.104151	5.592841	1.242424	0.488095
FX	27.83	2.969121	1.876173	1.166667	0.507246
GX	35.36	2.526848	1.557632	1.037736	0.395683
HX	34.95	3.703704	2.239642	1.034483	0.394737
JZ	59.82			0.409091	0.236842
JZ	50.96			1.7	0.566667
KZ	55.8			0.384615	0.2
LZ				0.4	0.2
M5	79.08	2.263724	1.328904	0.431818	0.223529
N5	85.46	2.023268	1.023541	0.470588	0.197531
O5	74.22	3.920646	2.311604	0.515152	0.220779
P5	74.93	0.969462	0.587889	0.333333	0.12963
Q5	93.78	4.088307	2.466091	0.486486	0.264706
R5	57.32	5.788712	3.095975	0.47619	0.222222
S5	44.05	7.320644	3.225806	0.352941	0.2
T5	30.64	0	0	0.32	0.190476
U5	17.63	0	0	0.416667	0.227273
V5	14.79	0	0	0	0
X5		0	0	0	0

attempts, increasing the likelihood of damage each time a tool is used, increasing the length of time a spear is kept on hand, and increasing the wait times at ambush locations. Finally, extending the duration of use would have an ultimate effect on the production and consumption of woodworking tools because spears are relatively costly to produce and maintain yet have a finite shelf life because they become brittle and more prone to catastrophic failure upon drying out (Gould 1980:129).

As a result, there appears to be a greater demand for spear production and maintenance when prey size decreases, diet breadth widens, and returns from hunting decrease. The result is consistent with the expectations of Kuhn and Miller's (2015) model, but this particular example

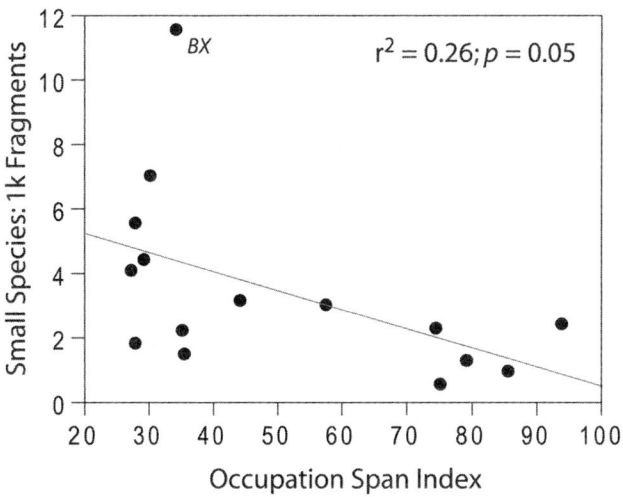

FIGURE 3.2. Small species diversity (number of small species taxa/1,000 bone fragments) compared to Surovell's occupation span index.

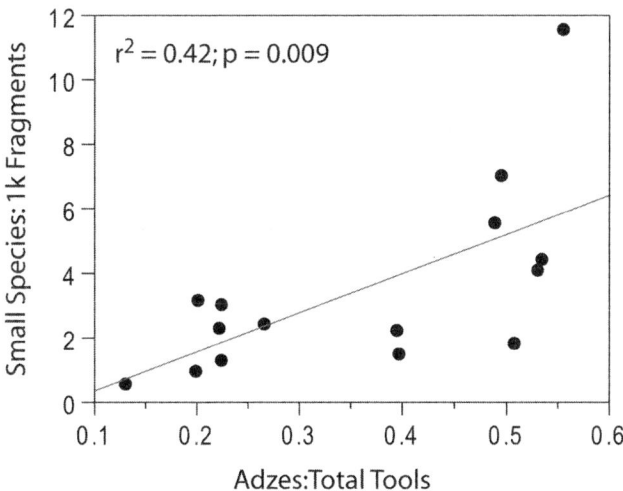

FIGURE 3.3. Small species diversity (number of small species taxa/1,000 bone fragments) compared to the frequency of woodworking tools (adzes/total tools).

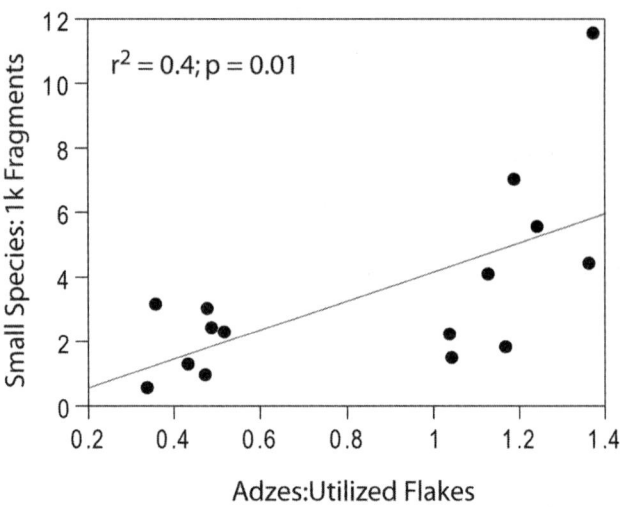

FIGURE 3.4. Small species diversity (number of small species taxa/1,000 bone fragments) compared to the frequency of woodworking tools (adzes/utilized flakes).

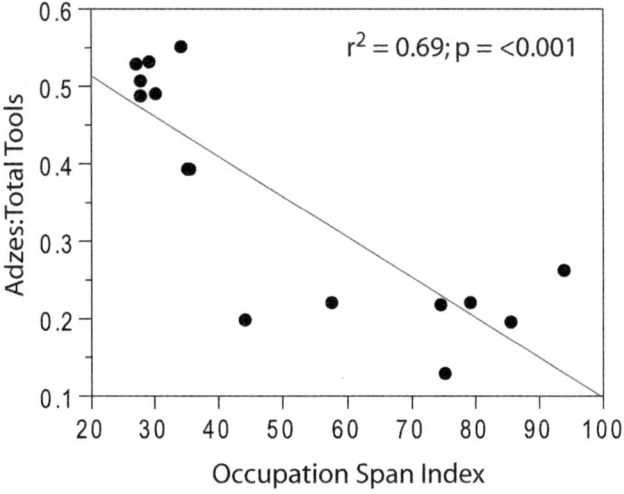

FIGURE 3.5. The frequency of woodworking tools (adzes/total tools) compared to Surovell's occupation span index by level.

really emphasizes the frequency of wooden spear maintenance. The following example will explore the relationship between prey size and the condition of discarded stone projectile points from an archaeological case study.

Gatecliff Shelter, Nevada

Gatecliff Shelter was initially discovered in 1970 by David Hurst Thomas in the Monitor Valley of Nevada and was subsequently excavated from 1971 to 1978 (Thomas 1983). A sequence of deposits 12 m deep was excavated, with over 650 m^3 removed. The site contained 56 distinct strata, including 16 cultural horizons that were dated with a total of 47 ^{14}C dates and a layer of temporally diagnostic Mazama ash. Gatecliff Shelter is also known for its exceptional faunal preservation, with over 51,000 animal bones recovered from the excavation (Grayson 1983). The remains recovered ranged in size from small mammals, such as rats and voles, to larger artiodactyl species, including pronghorn (*Antilocapra americana*) and bighorn sheep (*Ovis canadensis*). Over 400 projectile points were also recovered from this site, which help to provide the foundation for the regional chronology (Thomas and Bierwith 1983).

McGuire and Hildebrandt (2005) used Gatecliff Shelter as part of their argument that during the Middle to Late Holocene transition, prehistoric hunters in the Great Basin were acquiring large animals above and beyond what would be expected if they were simply economizing for calories. They interpret their sample of sites from across the Great Basin through the lens of costly signaling theory (e.g., Bliege Bird and Smith 2005) and argue that hunting to the point of being wasteful may have served as a prestige signal (McGuire and Hildebrandt 2005:708). To illustrate this quantitatively, they use the ratio of large mammals (i.e., pronghorn and bighorn sheep) to jackrabbits (*Lepus sp.*) and cottontails (*Sylvilagus sp.*) to show that during the Middle Archaic there was a spike in the representation of large species relative to small species.

Byers and Broughton (2004) and Byers and Smith (2007) argue that this increase in artiodactyls is likely due to ameliorating climate during the Late Holocene, where dramatic increases in artiodactyl remains are found in nonarchaeological assemblages, such as Homestead Cave (Byers and Broughton 2004:240–242). Byers and Broughton (2004) applied their version of the artiodactyl index (Σ Artiodactyl NISP/Σ Artiodactyl NISP + Σ Lagomorph NISP) to a sample of archaeological assemblages elsewhere in the Great Basin and found a dramatic increase in artiodactyl representation between 4,000 and 3,000 ^{14}C years ago. They argued that the frequency of burned lagomorph remains were constant over time and that there was no correlation between human and raptor-modified specimens, which indicated that the relative contribution of human and nonhuman predators varied little over time (Byers and Broughton 2004:244). They did, however, find a relationship between the artiodactyl index and the ratio of projectile points to cordage. They argued that as larger game became more abundant, there was an increase in the use of projectile points (presumably used for hunting artiodactyls) relative to nets, which were used in ethnographic contexts to hunt rabbits (Byers and Broughton 2004:245). They used these multiple lines of evidence to argue that during the Late Holocene, as artiodactyl populations increased, humans focused their diet breadth and preferentially targeted this species relative to smaller, harder-to-catch species.

Following both McGuire and Hildebrandt (2005) and Byers and Broughton (2004), I planned to use the ratio of artiodactyls to jackrabbits/cottontails as my proxy for prey size selection for each component at Gatecliff Shelter and then compare this ratio to the condition of the associated bifaces. However, Donald K. Grayson (1983:99) argued that the bones of many smaller species were likely brought into the cave by avian predators or wood rats (*Neotoma sp.*). As a result, the frequency of smaller species in each component may not reflect the prey size choices of human hunters if instead other agents are introducing them. As a means to test for the extent to which the small species representation is due to nonhuman predation, I compare the NISP of jackrabbits and cottontails to mice and voles, species for which there is no evidence for human consumption. This resulted in a clear, positive relationship ($r^2 = .79$; $p<0.0001$), which suggests that a significant percentage of the NISP can likely be attributed to factors other than human predators (Figure 3.6; Table 3.5). To minimize the effects of natural accumulation, I extract the residual values for each component from the previous regression and then take those residuals and regress them against the NISP of sheep and pronghorn. I use the residuals of this second regression analysis as a

TABLE 3.5. The NISP of mice, rats, and voles, and cottontails and jackrabbits by strata/horizon at Gatecliff Shelter, Nevada.

Strata/Horizon	Mice, Rats, and Voles	Cottontails and Jackrabbits	Residuals
Horizon I	131	129	−28.69143351
Horizon 2	370	526	73.14393119
Horizon 3	727	876	−17.75052405
Horizon 4	350	547	118.8439007
Horizon 5	179	362	145.0286397
Horizon 6	296	654	292.5338183
Strata 6/7	59	146	77.22845662
Stratum 8	25	31	4.218404737
Stratum 9	166	360	159.0836199
Stratum 10	5	23	20.91837422
Strata 11/12	212	210	−47.72630991
Stratum 13	104	19	−105.3464747
Strata 14–16	0	1	5.093366588
Stratum 17	12	14	3.273384899
Stratum 18	2	0	1.62336964
Stratum 19	150	54	−127.1564045
Stratum 20	14	18	4.803387951
Stratum 21	1	0	2.858368114
Stratum 22	34	30	−7.89658153
Stratum 23	26	14	−14.01659374
Strata 24/25	126	79	−72.51644114
Strata 26–30	59	8	−60.77154338
Strata 31/32	105	63	−62.58147319
Stratum 33	302	160	−208.8761726
Stratum 37	96	83	−31.46648692
Stratum 54	202	154	−91.37632517
Stratum 55	1	0	2.858368114
Stratum 56	428	332	−192.4859803

measure of the relative abundance of artiodactyls versus jackrabbits and cottontails (Figure 3.7; Table 3.6).

When compared to the biface assemblage, a negative trend occurs, where an increase in the relative abundance of smaller game yields a higher percentage of complete bifaces (Figure 3.8). The two Underdown components that drive this pattern are found in Horizon 2, which contains a bone bed with the remains of nearly two dozen adult bighorn sheep that were likely killed in a single event (Thomas and Mayer 1983:

TABLE 3.6. Hafted biface and faunal representation by horizon at Gatecliff Shelter, Nevada.

Horizon	Component	Bifaces (n)	Complete Bifaces (%)	Ovis/Antilocapra	Cottontails/ Jackrabbits	Residuals: Cottontails/ Jackrabbits	Ovis/Antilocapra: Adjusted Cottontail/ Jackrabbits
1	Yankee	19	68	18	129	−23.263	0.235
2	Underdown	22	55	500	526	80.974	2.763
3	Underdown	29	62	106	876	−6.331	1.132
4	Reveille	68	75	79	547	126.473	0.349
5	Reveille	106	76	68	362	150.939	0.271
6	Reveille	56	73	44	654	299.620	0.110
4–6	Reveille	230	75	191	1563	192.344	0.653
7	Reveille	39	69	45	146	81.933	0.247
8	Reveille/Devil's Gate			53	360	164.863	0.200
9	Devil's Gate	38	73	11	210	−41.484	0.188
10	Devil's Gate?	17	88	2	19	−100.190	−10.530
11	Devil's Gate?			0	2	7.505	0.000
12	Unknown	5	60	9	54	−121.537	−0.418
13	Clipper Gap	6	57	0	18	9.055	0.000
14	Clipper Gap	1	0	4	18	−3.444	0.041
15	Clipper Gap			2	30	−67.139	0.061
16	Clipper Gap			0	8	−56.067	0.000

355, 379). Moreover, Thomas and Mayer (1983:370) argue that the elements represented in the assemblage are consistent with an "ideal reverse utility strategy" (e.g., Binford 1978), where low-utility elements, such as mandibles, are discarded while those with higher economic utility are transported elsewhere. Compared to other components, the faunal record of Horizon 2 indicates a situation where the returns from hunting were exceptionally high, and a higher proportion of broken bifaces were discarded and not resharpened.

The high proportion of broken bifaces in the context of high returns from hunting is consistent with the application of the MVT to tool discard, where one of the less intuitively obvious predictions is that "the optimal point to abandon an artifact declines as average returns increase" (Kuhn and Miller 2015:181). In other words, as optimizing tool

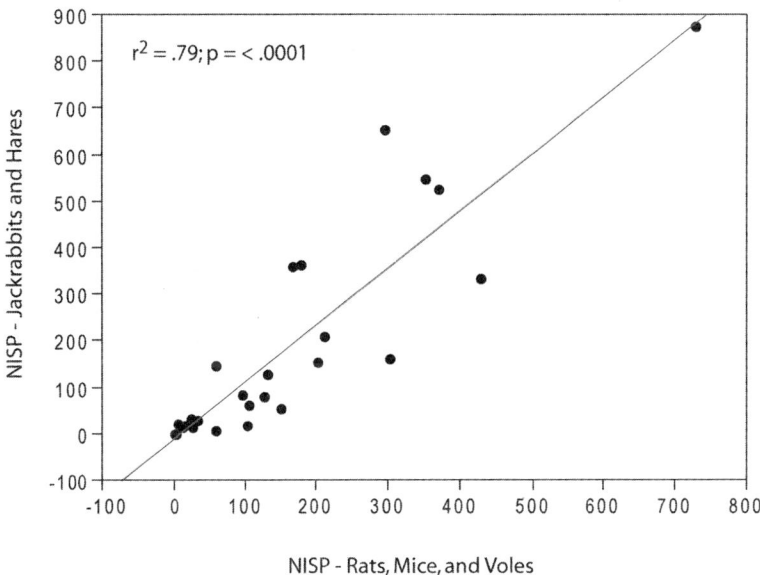

FIGURE 3.6. The NISP for rats, mice, and voles regressed against the NISP for jackrabbits and cottontails by horizon at Gatecliff Shelter, Nevada.

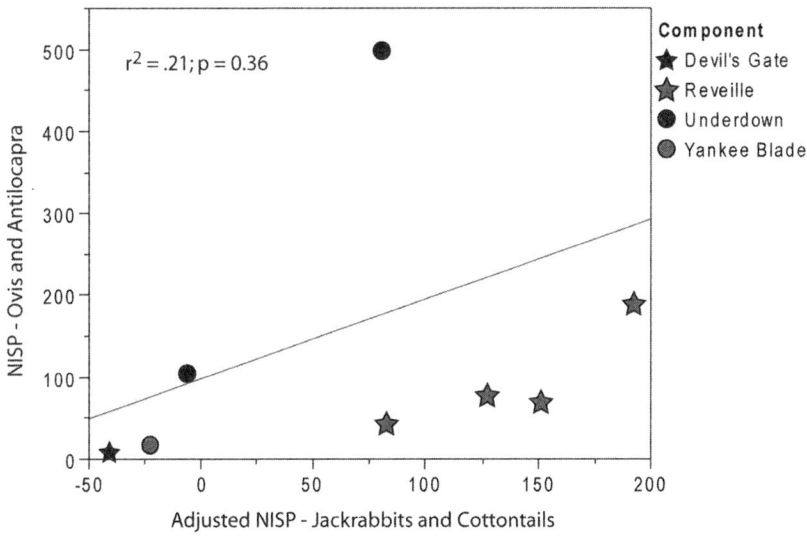

FIGURE 3.7. The adjusted NISP for jackrabbits and cottontails regressed against the NISP for antelope and bighorn sheep by component at Gatecliff Shelter, Nevada.

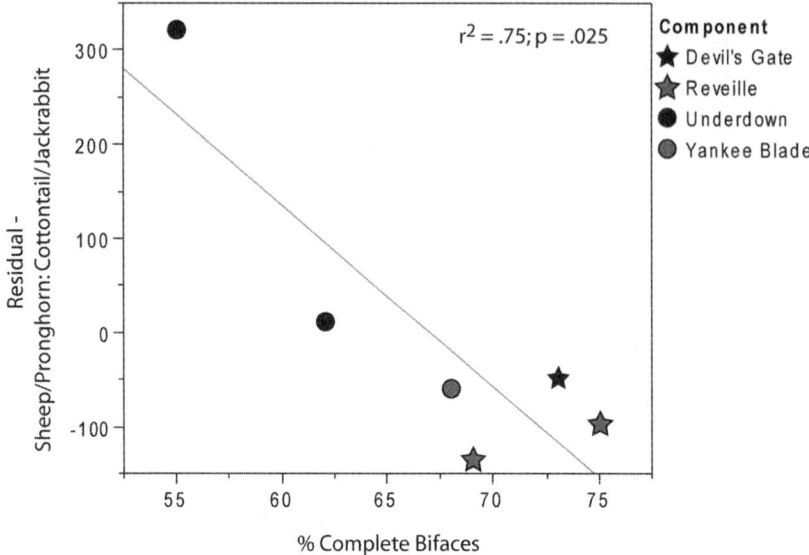

FIGURE 3.8. The percentage of complete bifaces regressed against the residuals from the regression illustrated in Figure 3.7.

users stand to gain more from the use of particular artifacts, they should replace them with new ones more often. In this case, the hunters of Horizon 2 from Gatecliff Shelter are similar to the caricature of the rifle-wielding hunter discussed earlier. Similarly, rather than carrying large amounts of ammunition and firing repeatedly, the hunters of Horizon 2 likely prepared and expended a relatively large number of spears and, as a result, left behind a high frequency of broken bifaces that could have otherwise been salvaged and reused.

Moreover, Kuhn and Miller (2015:181–184) state, "Conversely, if the average yield from the application of the tools goes down, people should hold on to artifacts longer." Or, to reframe this again using the rifle-wielding hunter example, when yields are lower, our hunter should practice more cautious shot selection and conserve ammunition. At Gatecliff Shelter, when returns were comparatively lower, a higher percentage of bifaces were discarded in their complete form. The high frequency of complete bifaces suggests frequent resharpening, and the specimens recovered were likely simply lost or reused past the point at which they were deemed useful and were subsequently replaced.

Conclusion

Bifaces are a critical artifacts class in North American archaeology because of their role as temporally diagnostic artifacts, which also make them a valuable point of departure for making inferences regarding human behavior. As a result, they have played a significant role in the development of the organization of technology theoretical perspective in Paleoindian and southeastern archaeology as well as recent attempts to apply more formal models from human behavioral ecology to stone stools.

As an example, I adopted Kuhn and Miller's (2015) artifacts-as-patches model to make tractable predictions that can then be applied to the archaeological record. One implication is that if variation in raw material quantity and quality are held constant, the strategies for discarding artifacts are likely related to prey size and, by extension, hunting returns. At Puntutjarpa Rockshelter in Australia, when prey size decreases and diet breadth widens, the frequency of tools used to maintain wooden spears increases. At Gatecliff Shelter, when hunting returns were exceptionally high, it appears that little effort was made to rejuvenate points for later use. Both of these findings were consistent with the expectations of Kuhn and Miller's (2015) model. With this model, and the chronological and environmental information presented in Chapter 2, it is now possible to construct a null model that can be used to make inferences about hunting returns from the first appearance of people to the emergence of domesticated plants in eastern North America.

CHAPTER 4

Projectiles Points and Prey Size
in the Lower Tennessee River Valley

Introduction

While the timing of the emergence of domesticated plants in eastern North America has been thoroughly documented, the role of environmental change, population pressure, and resource scarcity as causal factors is still heavily debated (Smith 2011, 2014; Gremillion et al. 2014a, 2014b, 2014c). In Chapter 1, I argue that the formal models of behavioral ecology provide an avenue to examine this issue. To do so, I advocate adopting the approach used by Stiner (2001), who used expectations from the diet breadth model, coupled with paleoenvironmental reconstructions of the Mediterranean Basin, to create a null model to compare against the archaeological record. For the Mid-South, I adopt a similar approach, but rather than examining changes in fauna, I argue that by modeling bifaces as patches of utility using the marginal value theorem (e.g., Kuhn and Miller 2015), it is possible to derive predictions for the archaeological record, and the bifaces in particular, if access to raw material can be held constant.

As a result, a series of predictions can now be made for how the biface inventory should change over time for the Mid-South. First, from the Bølling-Allerød through the Younger Dryas, there should be a noticeable shift in the condition of bifaces in the archaeological record with the extinction of megafauna. Kuhn and Miller's (2015) model predicts that since yields would decline in this context, there should be evidence for increasingly resharpened bifaces, where the use-life has been extended through resharpening. Using measurements reported in the Tennessee Fluted Point Survey, Kuhn and Miller found evidence for a substantial increase in resharpening during this time span, and the expectation is that this study should replicate those results.

Second, from the Younger Dryas through the mid-Holocene, temperatures in the Mid-South increased substantially and during the peak of the mid-Holocene became much drier. Gardner (1997) and, later, Thaddeus Bissett (2010) have argued that during the mid-Holocene, the interaction between temperature and moisture led to changes in forest structure, which fueled an increase in the abundance of deer. More specifically, as annual temperature increases, there is an increase in the frequency and abundance of mast production, which in modern deer populations have been linked to a decrease in birth spacing and an increase in the frequency of birthing of fawns as well as an increase in the survival rate of subadults. Additionally, Bissett (2010) finds that studies of modern deer populations have also shown that an increase in mast production increases the overall body mass of deer populations. Consequently, during the transition from the Early Holocene to the Middle Holocene, the expectation is that deer should be more prevalent and Kuhn and Miller (2015) would predict that bifaces should be replaced more frequently or show increasingly less evidence for resharpening to extend the use-lives of the bifaces.

Third, as temperature declined after the mid-Holocene, oak and hickory pollen diminished at Anderson and Jackson ponds (Delcourt 1979; Delcourt and Delcourt 1985). This likely signaled a decline in oak and hickory mast production compared to the peak temperatures of the mid-Holocene, which would have led to a decrease in the abundance of deer. Kuhn and Miller's (2015) model predicts that hunters at this time would have again sought to extend the use-lives of their bifaces.

In order to test these predictions, I used a sample of bifaces from Benton and Humphreys counties in the lower Tennessee River drainage. This group of collections contains a continuous sample of bifaces spanning the 10,000 years of prehistory, from the initial occupants of this area to the time at which the earliest domesticated seeds appear in the archaeological record.

Study Sample and Data Collection

Benton and Humphreys counties are located on either side of the lower Tennessee River and have been the focus of a considerable amount of research since the 1930s (Dye 2013). In particular, archaeologists working at the behest of the WPA conducted the first systematic survey of

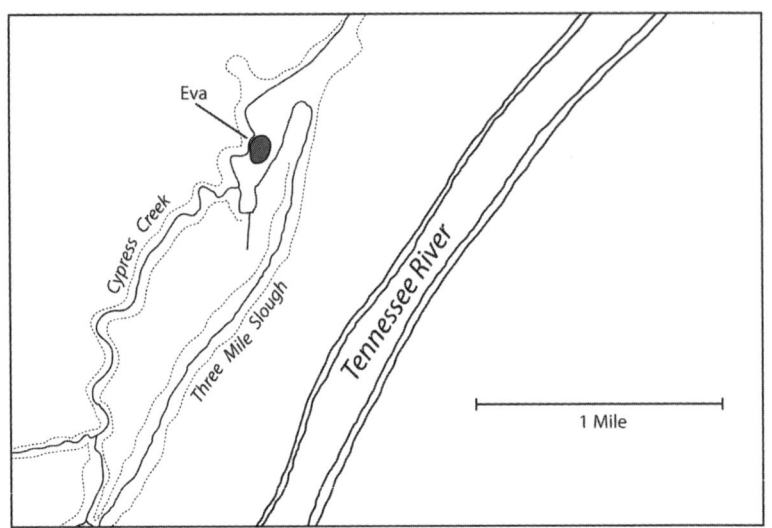

FIGURE 4.1. (*Top*) Location of the Eva site relative to the Tennessee River (modified from Lewis and Lewis 1961:ii). (*Bottom*) N-S Trench at the Eva site (McClung Museum of Natural History and Culture WPA/TVA Archaeological Photo Archives).

this drainage in advance of the construction of Kentucky Lake and subsequently excavated a series of sites that became the foundation for the regional culture-historical sequence for the Mid-South (Lewis and Kneberg 1959).

Part of the success that the WPA archaeologists had in locating, and then excavating, a large number of sites is due to alluvial geomorphology in this section of the river drainage. Here the downcutting is constrained by the bedrock geology, so this section of the river is characterized by a relatively high degree of lateral movement. The lateral movement produced a series of levees radiating away from the main channel of the river (Leach and Jackson 1987). The levee system is beneficial for archaeological site discovery and excavation because it created the context for burial in levee deposits, while the lateral movement of the river prevented sites from becoming deeply buried. For example, the Eva site is located on a levee 1.2 km away from the main channel, and the artifact-bearing layers only extended ~2 m below the surface and contained Early Archaic through Woodland period deposits (Figure 4.1).

The subsequent inundation of Kentucky Lake, and the seasonal rising and falling of the lake, has eroded and exposed many of the archaeological sites in these counties. In some areas, this has completely exposed and deflated Paleoindian and Early Archaic sites located further away from the main channel of the river, which have since been collected by both professional (Broster and Norton 1996; Kerr and Bradbury 1998; Lewis and Kneberg 1958) and avocational archaeologists (McNutt et al. 2008) in the decades after the creation of Kentucky Lake. Important Paleoindian and Early Archaic sites include Nuckolls (Lewis and Kneberg 1958; Norton and Broster 1992), Twelkemeier (Broster and Norton 1990), and Carson-Conn-Short, where Broster and Norton (1993, 1996) have reported a limited area of intact, buried deposits. In other cases, most of the available information on sites comes from private collections. For example, Ernie Simms donated his entire collection to the McClung Museum of Natural History and Culture, which includes the primary assemblage from the Kirk Point site (McNutt et al. 2008). In other cases, information from private collections was incorporated directly into the Tennessee Fluted Point Survey (Broster 1989; Broster and Norton 1996;

Broster et al. 2013). Finally, since Kentucky Lake is managed jointly by the Tennessee Valley Authority and the Tennessee Wildlife Resources Agency, it has been surveyed multiple times, including a large-scale survey by Cultural Resource Analysts Inc.(CRA) (Kerr 1996). As a result of these various projects, Benton and Humphreys counties contain one of the most abundant and temporally complete inventories of artifacts—in particular, bifaces—available for study in all of eastern North America.

As a result, the biface assemblages of Benton and Humphreys counties are ideal for two reasons. First, there are abundant raw material sources (Amick 1987). The Tennessee and Cumberland drainages in central Tennessee contain multiple chert-bearing limestone formations that provide ready sources of raw material for stone tools. The assemblages consist primarily of Dover and Fort Payne chert types. Dover chert occurs at the contact of the St. Louis and Warsaw formations and is found near Dover, Tennessee, although nodules are frequently found along the Tennessee River in the northern parts of Benton and Humphreys counties (Parish 2011). Fort Payne Formation chert—in particular, the variety known locally as Bulls Eye or Buffalo River chert—is found near the confluence of the Tennessee and Duck rivers at the southern part of the study area. In addition, other chert types are likely to be found from outside the research area, most likely from the Nashville Basin or the St. Louis and St. Genevieve formations located in the Cumberland River drainage.

The abundance of lithic raw material in this area provides an archaeological context where access to raw material can be held constant. Probably at no point in prehistory was a hunter-gatherer more than a single day's foray away from a source of high-quality raw material. As a result, variation in resharpening likely reflects factors other than raw material scarcity, including fluctuations in hunting returns from the Pleistocene through the mid-Holocene.

Second, there are many collections available for study from this area (Figure 4.2). This includes one excavated Early Paleoindian period Clovis site, Carson-Conn-Short (40BN190) (Broster and Norton 1993, 1996), which is located at the confluence of the Tennessee and Duck rivers. Two additional sites contain Paleoindian period artifacts, Nuckolls (Broster and Norton 1991; Lewis and Kneberg 1958; Norton and Broster 1992) and Twelkemeier (Broster and Norton 1990). Both sites have been ex-

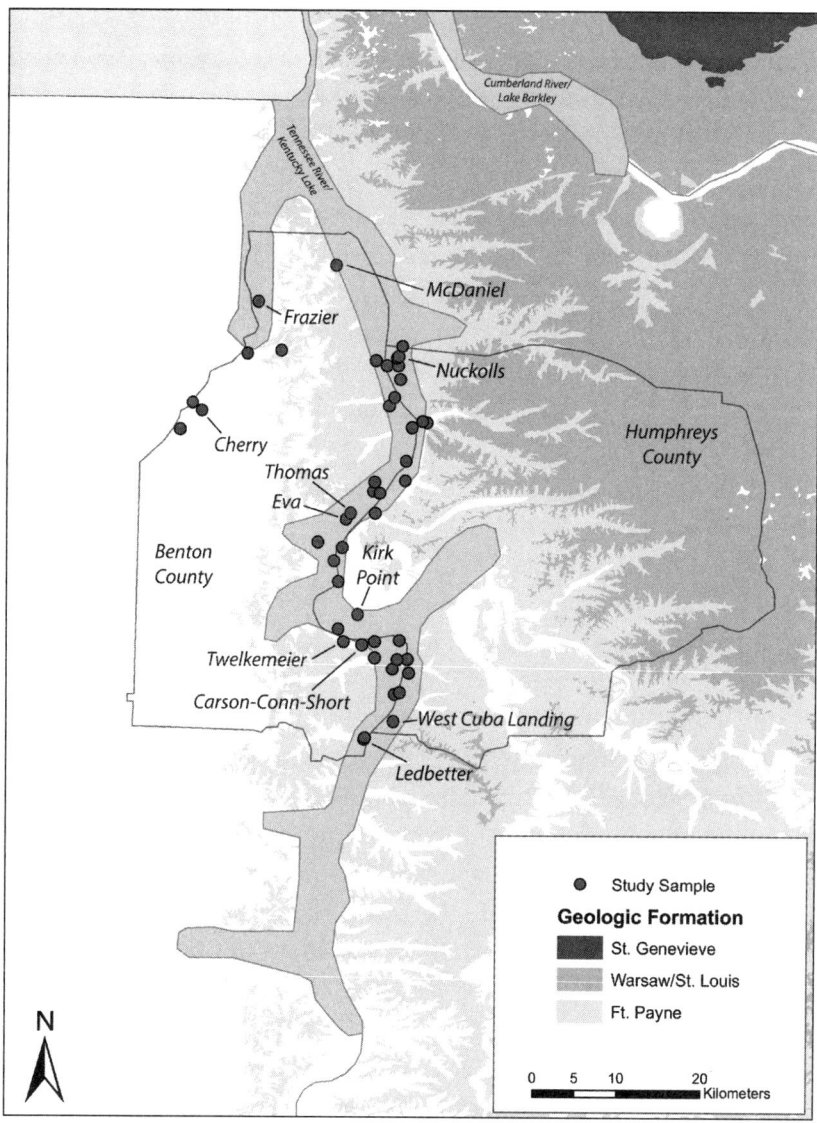

FIGURE 4.2. Study area, site locations, and the distribution of chert-producing St. Genevieve, Warsaw/St. Louis, and Ft. Payne limestone formations.

posed and deflated by the periodic raising and lowering of Kentucky Lake. However, large collections from both sites are curated at the Tennessee Division of Archaeology (TDOA) and the McClung Museum at the University of Tennessee.

As for post-Paleoindian occupations, several sites from this section of the lower Tennessee River valley excavated as WPA projects provided the basis for Lewis and Kneberg's cultural sequence for the Archaic period for the mid-continent (Lewis and Kneberg 1959). These include the Middle and Late Archaic sites of Eva (Lewis and Lewis 1961), West Cuba Landing (Lewis and Kneberg 1958), Ledbetter (Higgins 1982), Thomas (Lewis and Kneberg 1958), Frazier (Lewis and Kneberg 1958), Cherry (Magennis 1977), and McDaniel (Lewis and Kneberg 1958). Artifacts collected during the course of a shoreline survey of Kentucky Lake by CRA (Kerr 1996; Kerr and Bradbury 1998) are also included in this study.

One drawback to using this sample of assemblages is that, aside from Carson-Conn-Short, no artifacts except obvious shaped tools were collected. In the case of Nuckolls and Twelkemeier, this was because these sites were deflated surface scatters containing multiple cultural components. For the sites excavated as part of the WPA, lithic debitage and unmodified faunal remains were not curated (Bissett 2014; Chapman 1988). However, it appears that almost all of the bifaces recovered from each of the sites in the study sample were collected and curated.

In December 2010 and July and August 2011, I collected the primary information used in this analysis. Data collection was a multistep process that first began with taking digital photos of each artifact from a fixed elevation using a Canon EOS Rebel DSLR camera and recording its thickness with digital calipers. I used Adobe Photoshop CS 5.5 to crop each artifact image, scale it to the actual size of the artifact, and then reset the canvas size so that each image had the same overall dimensions. The artifact images for each site were then combined into a single document in Adobe InDesign to create a virtual table of artifacts, which was also exported as an image file (.tiff). I imported these image files into Esri ArcGIS, where I converted each raster into individual polygons for each biface. The Envelope function was then used to generate a maximum bounding box around each artifact, which provided a means to calculate the maximum width and length. These measurements, as well as image files for each artifact, were then imported into Microsoft Access, where they were integrated into a relational database that also includes the initial caliper measurements.

Within Microsoft Access, I also created a coding sheet to record qualitative attributes, primarily reflecting the condition of the artifact and its typological classification. From this point, I was able to integrate the artifacts recorded in the CRA survey of Kentucky Lake as well as the bifaces reported in the Tennessee Fluted Point Survey from Benton and Humphreys counties. This provides me with an overall total of 5,244 bifaces from 87 sites from within my study area.[1]

Because of the relatively large sample size, I could afford to be selective in the artifacts I chose to analyze. First, I am primarily interested in the points that were produced for, and presumably used in, the context of hunting. Consequently, points that were intentionally placed within burials were likely transferred from the systemic to the archaeological context (e.g., Schiffer 1983) prematurely, meaning that they were discarded before their utility was completely expended (Shott and Ballenger 2007:156). Consequently, points from clear burial contexts were excluded from the analysis.

Also, for some point types, like the Big Sandy side-notched and corner-notched points, there are both early and late forms that are difficult to separate based on visual inspection. For the Big Sandy Side-Notched, I only use the bifaces found in Stratum II at Eva, which is clearly Late Archaic in age. Similarly, for the corner-notched bifaces, I only use the points recovered at Nuckolls and Kirk Point, which have relatively large percentages of bifaces that are Paleoindian and Early Archaic in age, and any corner-notched points found at these sites are also likely to be Early Archaic in age rather than dating to the Woodland period.

The large sample size also allows me to be selective in regard to the measurements. The overwhelming majority of the points in my sample are not complete. However, for most basic measurements, I have robust sample sizes for each type in the study sample (Table 4.1; Table 4.2).

Methods

Following Kuhn and Miller (2015), I adopt a series of proxy measurements to describe how the size and shape of the bifaces change through time. Hunzicker (2008:306) simply used length as the y-axis in his attrition curve (Figure 4.3). This is essentially a measure of the size of the projectile point or, alternatively, a measure of the unexpended utility (e.g., Shott 1996:267). Hunzicker (2008) could use length because he was using replicas of Folsom points, which all share a similar form. In the sample of bifaces from the lower Tennessee River valley, form varies greatly over

TABLE 4.1. Archaeological components present at each site included in the study sample.

Site Number	Clovis	Cumberland	Quad/Beaver Lake	Dalton	Corner-Notched	Early Archaic Stemmed	Eva I	Eva II/Morrow Mountain	Middle Archaic Stemmed	Late Archaic Stemmed	Late Archaic Barbed
40BN0005							X	X		X	
40BN0006										X	X
40BN0007						X				X	
40BN0012						X	X	X	X	X	X
40BN0014						X			X	X	
40BN0017								X	X	X	X
40BN0018									X		
40BN0024										X	
40BN0025							X		X	X	X
40BN0028										X	X
40BN0029										X	
40BN0035										X	
40BN0039						X		X		X	X
40BN0047										X	
40BN0056						X				X	
40BN0058						X					X
40BN0059			X			X	X	X		X	X
40BN0059	X					X	X	X	X	X	X
40BN0074						X	X	X	X	X	X
40BN0077				X		X	X	X	X	X	X
40BN0081									X	X	
40BN0086										X	
40BN0088										X	

40BN0090
40BN0094
40BN0100
40BN0101
40BN0113
40BN0114
40BN0120
40BN0124
40BN0141
40BN0142
40BN0143
40BN0144
40BN0145
40BN0147
40BN0149
40BN0179
40BN0190
40BN0295
40BN0317
40HS0009
40HS0023
40HS0048
40HS0059
40HS0060
40HS0063
40HS0075

TABLE 4.1. (cont'd.) Archaeological components present at each site included in the study sample.

Site Number	Clovis	Cumberland	Quad/Beaver Lake	Dalton	Corner-Notched	Early Archaic Stemmed	Eva I	Eva II/Morrow Mountain	Middle Archaic Stemmed	Late Archaic Stemmed	Late Archaic Barbed
40HS0098			X								
40HS0173	X	X	X								
40HS0174	X	X	X	X							
40HS0175	X	X	X								
40HS0176	X	X	X								
40HS0181	X	X	X								
40HS0182	X		X								
40HS0183	X	X	X								
40HS0185	X	X	X								
40HS0186		X									
40HS0200	X	X	X								
40HS0205	X		X	X							
40HS0284	X	X	X	X							
40HS0333			X								
40HS0334	X										

TABLE 4.2. Sample size for each measurement by biface group.

Type	Thickness	Area	L:W[1]	L:T[2]	W:T[3]
Clovis	136	94	94	58	91
Cumberland/Barnes	78	62	62	48	63
Quad/Beaver Lake	220	90	90	66	178
Dalton	161	56	56	57	125
Corner-Notched	92	104	104	39	77
Early Archaic Stemmed	113	109	109	63	111
Eva I	203	62	62	78	164
Eva II/Morrow Mountain	89	46	46	52	79
Middle Archaic Stemmed	214	105	105	96	200
Late Archaic Stemmed	956	378	378	319	932
Late Archaic Barbed	153	85	85	76	143

[1] Maximum Length:Maximum Width
[2] Maximum Length:Maximum Thickness
[3] Maximum Width:Maximum Thickness

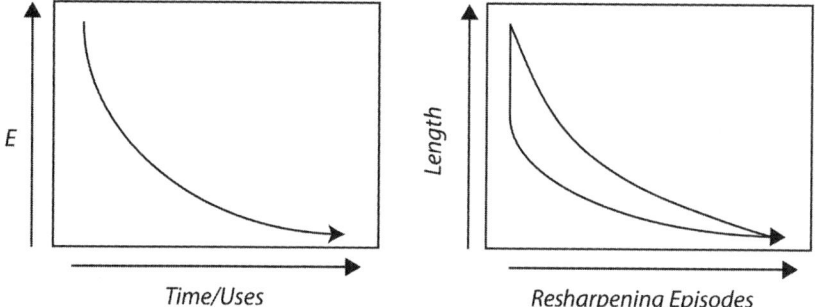

FIGURE 4.3. (*Left*) The relationship between *E* (the average expected instantaneous returns minus the cost of producing or procuring a new artifact) and time and/or the number of uses (modified from Kuhn and Miller 2015). (*Right*) The relationship between time and resharpening episodes for Folsom type bifaces fired and resharpened in an experimental context (modified from Hunzicker 2008:306).

time in some cases. As a result, other proxies for the absolute size of the point are required. Since I recorded the maximum length, width, and thickness for each biface, I can calculate the area and volume for each point instead.

Also, unlike Hunzicker (2008), there is no way to identify the number of uses, or the time in use, directly. Instead, I have to make the assumption that the degree of resharpening nonlinearly correlates with time in use, which, based on Hunzicker's experimental data, is a reasonable assumption. Following Kuhn and Miller (2015), I am able to calculate the length:width. This proxy assumes that with each resharpening episode, length will be differentially removed relative to the width, which is constrained by the portion that is within the haft. As a result, in situations where there is minimal resharpening, the size of the points will be free to vary, but their aggregate lengths and widths will be correlated. On the other hand, points that have been subjected to varying degrees of resharpening will not have lengths and widths that correlate in the same manner, because the length will be reduced at a much greater rate than the width with each resharpening episode (Figure 4.4).

Since many of the bifaces in my sample are incomplete, which makes it impossible to accurately assess maximum width and length measurement, I utilize two additional proxies for resharpening that can be applied to a larger number of bifaces in my sample, which are the length:thickness and width:thickness ratios. In both instances, maximum thickness is constrained by the portion of the biface that is located in the haft and is not affected much by resharpening, whereas resharpening will impact the length and width of the exposed portion of the point (Figures 4.5 and 4.6; Table 4.3).

For this analysis I examined thickness, area, length:width, length:thickness, and width:thickness using, first, analysis of variance (ANOVA) (Fisher 1918) to determine if the means across all of the types are statistically different. I then used the Tukey-Kramer method (Kramer 1956; Tukey 1953) to determine which pairs of means for each group of bifaces are statistically similar or different. Finally, I used principal component analysis (PCA) (Pearson 1901) to examine the combined effects and interactions of all of these variables through time.

Results

For the maximum thickness, the ANOVA test was statistically significant ($F = 146.49$, $df = 10$, $p<.0001$), which indicates that the means across all of the groups of bifaces are not the same (Figure 4.7). There appears to be a slight decrease in mean thickness during the Paleoindian period that is not statistically significant other than the difference in means between the sample of Clovis bifaces and the Quad/Beaver Lake and Dalton

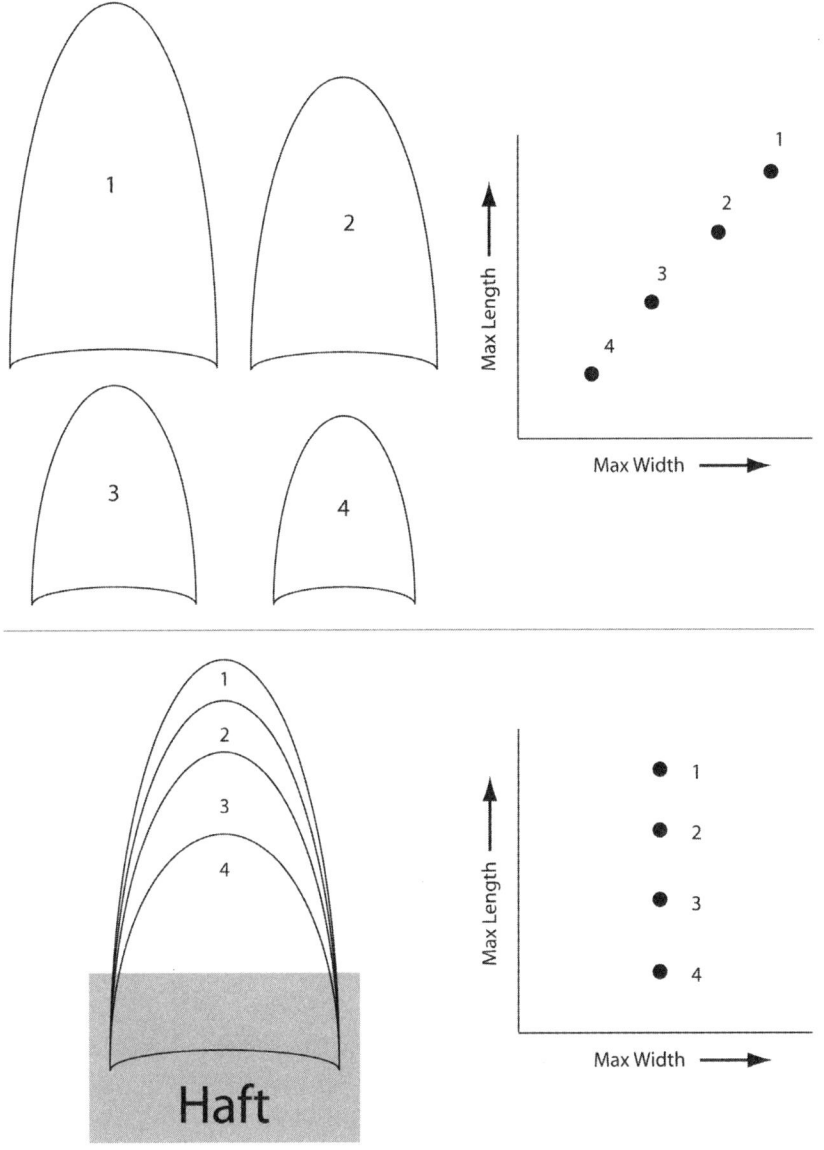

FIGURE 4.4. The hypothetical relationship between biface length, width, and resharpening.

TABLE 4.3. Means and standard deviations for each variable by biface group.

Type	Thickness			Area			L:W[1]			L:T[2]			W:T[3]		
	n	μ	σ	n	μ	σ	n	μ	σ	n	μ	σ	n	μ	σ
Clovis	136	7.59	2.78	94	22.54	0.80	94	2.55	0.06	58	10.33	3.98	91	4.19	0.08
Cumberland/Barnes	78	7.35	2.22	62	20.91	0.99	62	2.53	0.07	48	8.96	2.85	63	3.87	0.09
Quad/Beaver Lake	220	6.75	1.28	90	17.61	0.82	90	2.50	0.06	66	9.44	3.16	178	3.92	0.05
Dalton	161	7.21	1.14	56	13.91	1.04	56	2.26	0.07	57	7.82	2.01	125	4.15	0.07
Corner-Notched	92	7.70	1.07	104	17.22	0.76	104	1.89	0.05	39	6.63	1.15	77	3.96	0.08
Early Archaic Stemmed	113	7.84	1.24	109	16.77	0.74	109	2.04	0.05	63	7.37	1.52	111	3.84	0.07
Eva I	203	8.75	1.32	62	25.15	0.99	62	1.62	0.07	78	7.51	1.51	164	4.48	0.06
Eva II/Morrow Mountain	89	8.45	1.75	46	20.37	1.14	46	1.93	0.08	52	7.56	1.75	79	4.06	0.08
Middle Archaic Stemmed	214	8.14	1.47	105	16.60	0.76	105	2.04	0.05	96	7.33	1.90	200	3.66	0.05
Late Archaic Stemmed	956	10.48	2.01	378	23.81	0.40	378	2.22	0.03	319	7.13	1.67	932	3.43	0.02
Late Archaic Barbed	153	9.43	1.93	85	20.88	0.84	85	1.79	0.06	76	6.73	1.31	143	3.78	0.06

[1] Maximum Length:Maximum Width
[2] Maximum Length:Maximum Thickness
[3] Maximum Width:Maximum Thickness

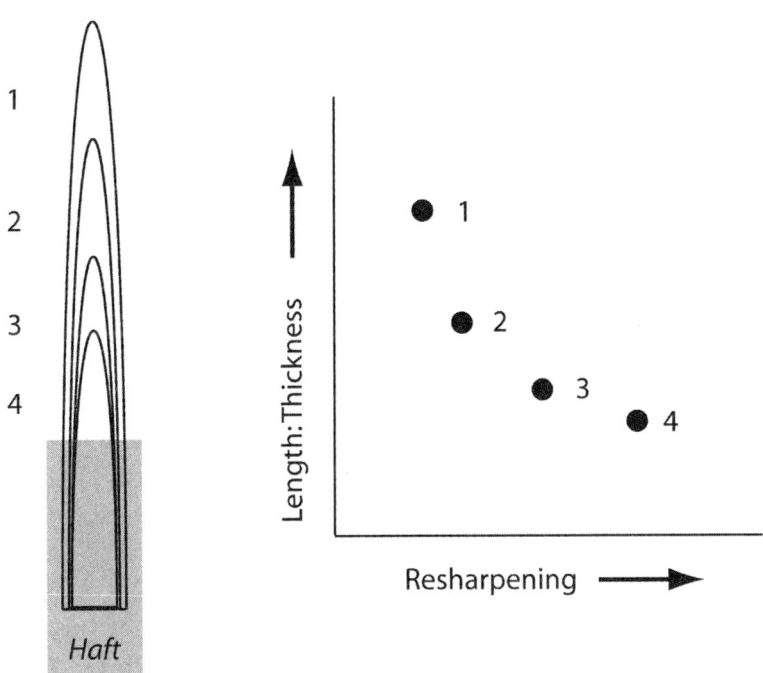

FIGURE 4.5. The hypothetical relationship between biface length, thickness, and resharpening.

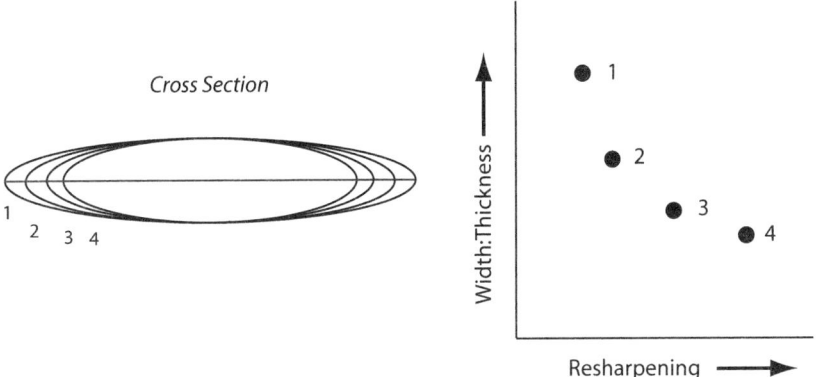

FIGURE 4.6. The hypothetical relationship between biface width, thickness, and resharpening.

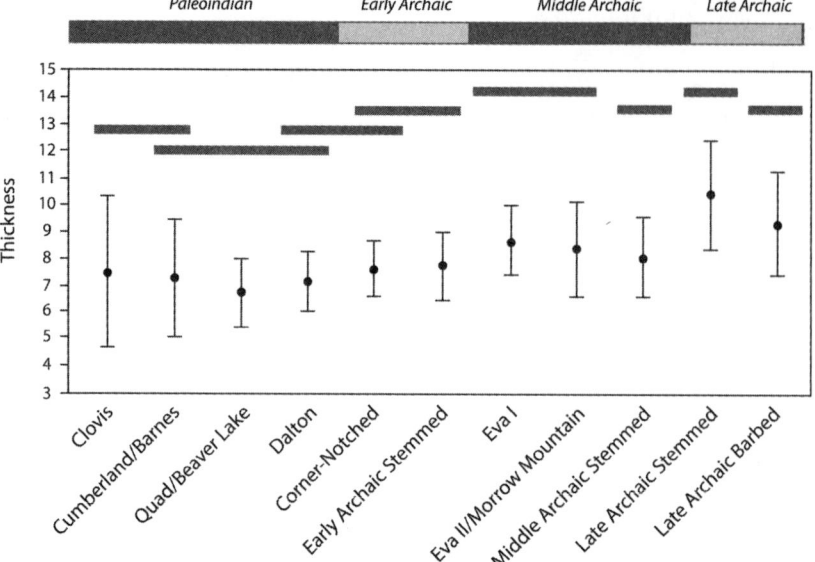

FIGURE 4.7. Means and standard deviations for the maximum thickness values for each biface group. Gray bars indicate whether the means of adjacent groups are statistically the same based on Tukey's HSD test.

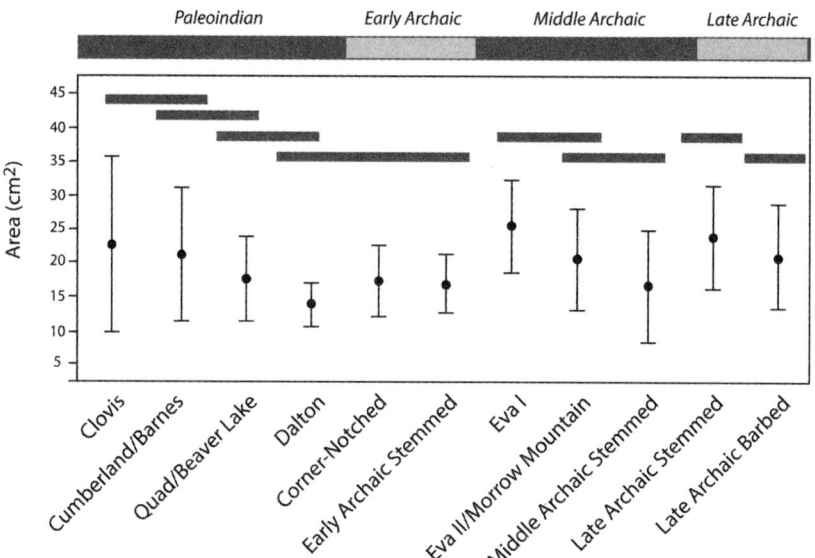

FIGURE 4.8. Means and standard deviations for the area values for each biface group. Gray bars indicate whether the means of adjacent groups are statistically the same based on Tukey's HSD test.

samples. However, there is a very noticeable decrease in the size of the standard deviations over the Paleoindian period. Then, over the course of the Archaic period, the means increase significantly, with the highest means values occurring with the Eva I, Late Archaic Stemmed, and Late Archaic Barbed groups.

To calculate the two-dimensional area for each point, I simply multiplied the maximum length by the maximum width for each point. This provides a value for the modular area of the point, or the size of the smallest rectangle that will contain it. An ANOVA test found significant differences in the mean values ($F = 22.15$, $df = 2$, $p<.0001$) (Figure 4.8). The means and standard deviations decrease over the course of the Paleoindian and Early Archaic periods and then increase significantly with the Eva I sample. They then decrease with the Eva II/Morrow Mountain and Middle Archaic Stemmed groups. This is followed by an increase with the Late Archaic Stemmed bifaces and then a decrease with the Late Archaic Barbed samples.

The ANOVA test for the length:width for all of the samples was statistically significant ($F = 26.33$, $df = 10$, $p<.0001$) (Figure 4.9). There is a slight decrease in the means and standard deviations in the Paleoindian samples that is not statistically significant except for the difference between the mean values for Clovis and Dalton. The mean values for the Early Archaic Corner-Notched and Early Archaic Stemmed groups were statistically the same but also significantly lower than the Paleoindian samples. Finally, the Eva I group had the lowest mean value, which is followed by a significant increase over the rest of the Middle and Late Archaic samples, with the Late Archaic Barbed sample being the lone exception.

For the length:thickness, the ANOVA test was statistically significant ($F = 21.44$, $df = 10$, $p<.0001$), which was driven largely by a significant decrease in both the means and the standard deviation from the Clovis through Corner-Notched samples (Figure 4.10). However, the results of the Tukey-Kramer test show that the mean values, as well as the standard deviations, stayed remarkably constant over the course of the Early, Middle, and Late Archaic periods.

The mean values for the width:thickness were also significantly different ($F = 45.74$, $df = 10$, $p<.0001$), and there appeared to be a slight

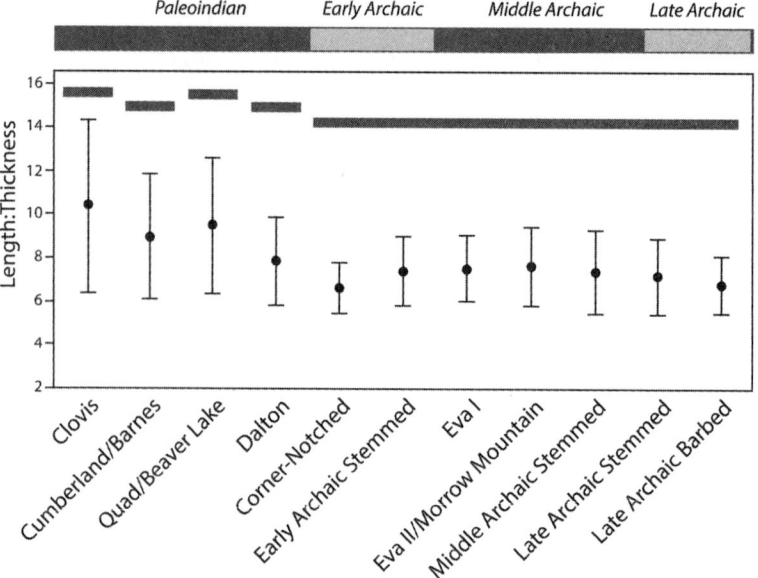

FIGURE 4.9. Means and standard deviations for length:width values for each biface group. Gray bars indicate whether the means of adjacent groups are statistically the same based on Tukey's HSD test.

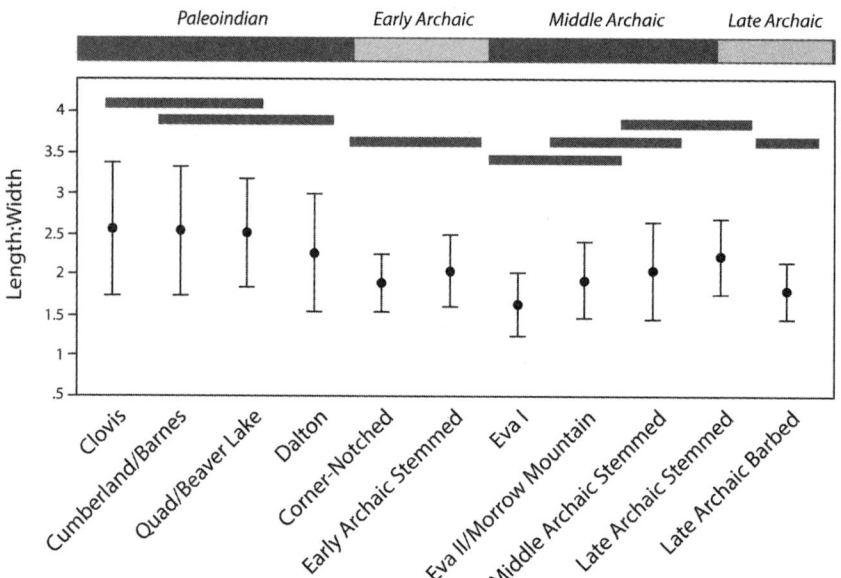

FIGURE 4.10. Means and standard deviations for length:thickness values for each biface group. Gray bars indicate whether means of adjacent groups are statistically the same based on Tukey's HSD test.

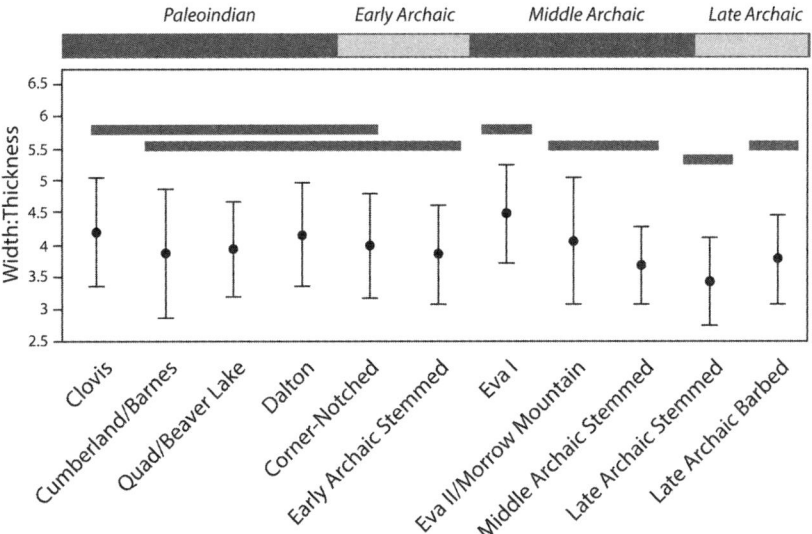

FIGURE 4.11. Means and standard deviations for width:thickness values for each biface group. Gray bars indicate whether the means of adjacent groups are statistically the same based on Tukey's HSD test.

decrease in mean values over the course of the Paleoindian and Early Archaic periods (Figure 4.11). However, the results of the Tukey-Kramer test indicate that this trend is not significant other than the difference between the means of the Clovis and Early Archaic Stemmed samples. There was a significant increase with the Eva I sample, followed by a decrease with the Late Archaic Stemmed sample. This was followed by a significant increase with the Late Archaic Barbed sample.

Finally, I generated a correlation matrix and conducted a principal component analysis (e.g., Pearson 1901) to examine how these various proxies for size and shape interact (Figures 4.12 and 4.13; Table 4.4). For the PCA, I used the correlation matrix for the basis of comparison since I used variables for both measurements (maximum width, maximum length, thickness, and area) and ratios (length:width, width:thickness, length:thickness), and thus the units of measurements were different between the variables. The correlation matrix descales the data, whereas the covariance matrix does not. For the analysis, the eigenvalues of the first three principal components account for 98.6 percent of the variation. The first eigenvector reflects variation in the maximum length and area variables, which are highly correlated (r = .85).

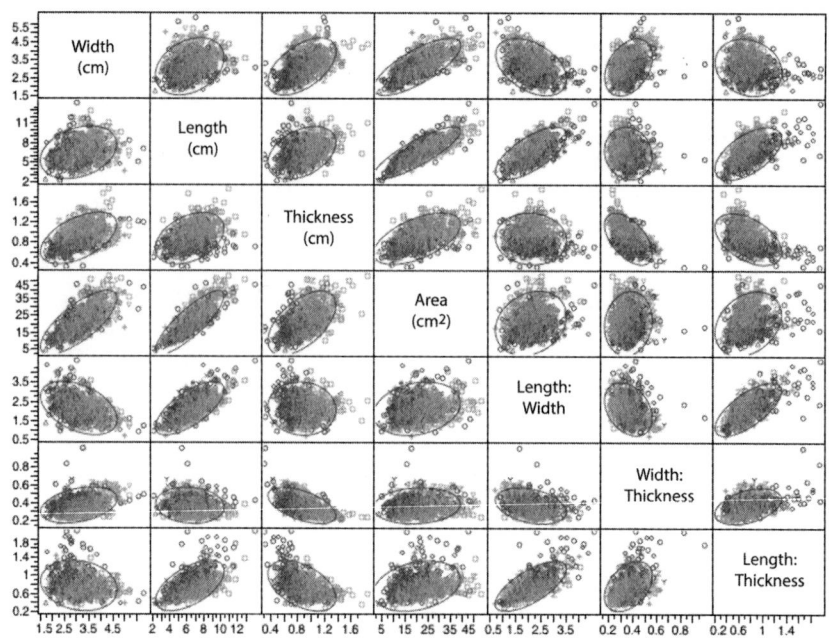

FIGURE 4.12. Correlation matrix for all of the variables used in this study.

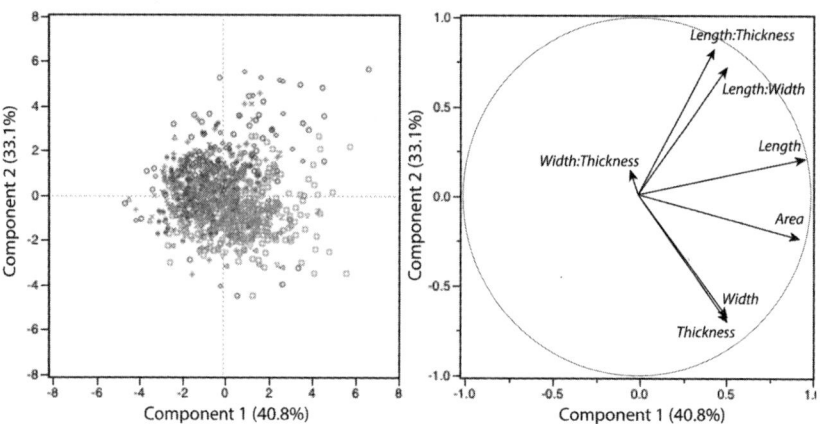

FIGURE 4.13. Principal components for all of the variables used in this study.

TABLE 4.4. The eigenvectors from the principal components analysis.

Variable	PC1	PC2	PC3	PC4	PC5	PC6	PC7
Max. Width (cm)	0.31	−0.45	0.38	−0.07	0.68	−0.24	−0.16
Max. Length (cm)	0.57	0.14	−0.07	−0.19	−0.24	0.22	−0.71
Thickness (cm)	0.31	−0.47	−0.33	0.74	−0.15	0.06	0.07
Area	0.56	−0.16	0.15	−0.39	−0.33	0.00	0.62
Length:Width	0.31	0.48	−0.32	0.09	0.57	0.39	0.28
Length:Thickness	0.27	0.54	0.24	0.37	−0.09	−0.66	0.03
Width:Thickness	−0.03	0.10	0.75	0.34	−0.10	0.55	0.04

Consequently, the first principal component is essentially variation in size, especially as it relates to length. The second principal component is driven by the length:width and length:thickness variables and their negative loading with the maximum width and thickness. This principal component is a proxy for the relative shape of the artifact, or the relative relationship between length, width, and thickness. It is basically a proxy for the relative degree of elongation or refinement. Finally, the third principal component is related to the variation in the width:thickness variable. I then saved the principal component values for each artifact and examined the variation for each biface group through time (Figures 4.14–4.16; Table 4.5).

For the first principal component, there appears to be a decrease in both the means and standard deviations from the Clovis through Early Archaic Stemmed groups, which indicates decreasing size from the Paleoindian period and into the Early Archaic. Then, with the Eva I group, the values become significantly larger and then increase for the remainder of the Middle Archaic. Finally, the values increase, and then decrease, with the Late Archaic Stemmed and Late Archaic Barbed types.

For the second principal component, during the Paleoindian period, there was a slight decrease in the mean values, but only the difference between Clovis and Dalton was significant. This trend continues throughout the beginning of the Middle Archaic with Eva I, albeit with smaller standard deviations than their Paleoindian counterparts. This trend indicates decreasing length relative to width and thickness and may reflect proportionality indicative of an increasing frequency of resharpening episodes. This trend then reverses itself during the remainder of the

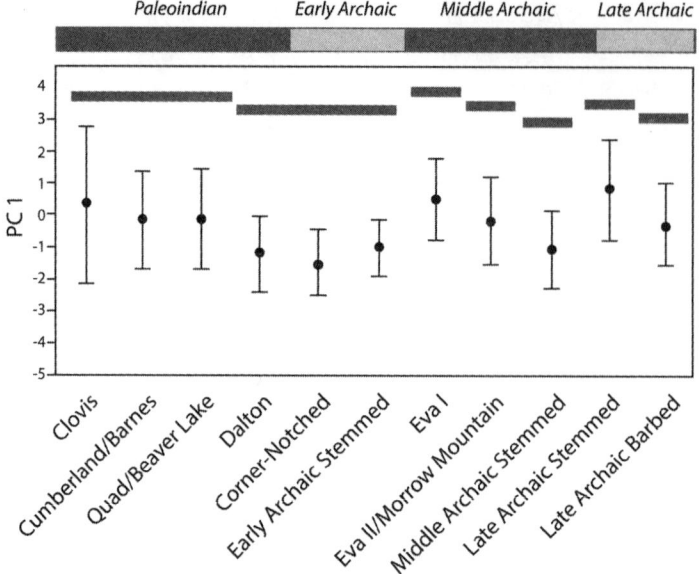

FIGURE 4.14. Means and standard deviations for the first principal component values for each biface group. Gray bars indicate whether the means of adjacent groups are statistically the same based on Tukey's HSD test.

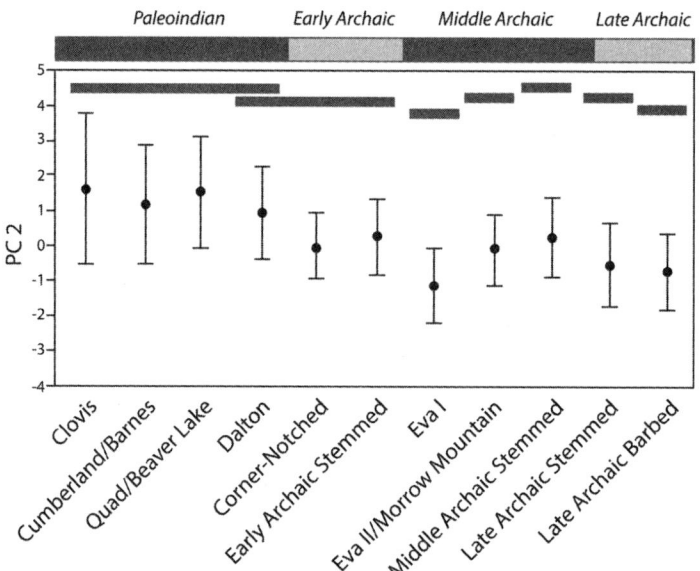

FIGURE 4.15. Means and standard deviations for the second principal component values for each biface group. Gray bars indicate whether the means of adjacent groups are statistically the same based on Tukey's HSD test.

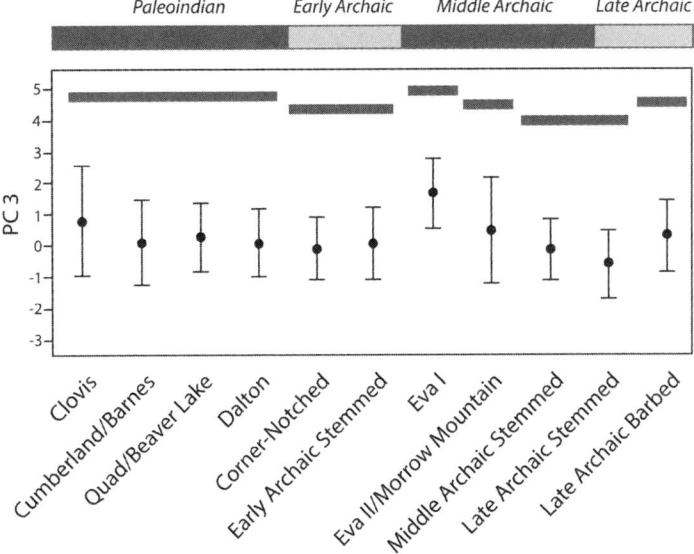

FIGURE 4.16. Means and standard deviations for the third principal component values for each biface group. Gray bars indicate whether the means of adjacent groups are statistically the same based on Tukey's HSD test.

TABLE 4.5. Means and standard deviations from the PCA values by biface group.

Type	PC1		PC2		PC3	
	μ	σ	μ	σ	μ	σ
Clovis	0.33	0.20	1.61	0.17	0.80	0.16
Cumberland/Barnes	−0.15	0.24	1.16	0.20	0.07	0.19
Quad/Beaver Lake	−0.18	0.19	1.48	0.16	0.24	0.15
Dalton	−1.19	0.25	0.92	0.21	0.07	0.19
Corner-Notched	−1.51	0.25	0.02	0.22	−0.10	0.20
Early Archaic Stemmed	−1.04	0.19	0.24	0.17	0.06	0.15
Eva I	0.50	0.20	−1.14	0.17	1.65	0.16
Eva II/Morrow Mountain	−0.21	0.22	−0.13	0.19	0.48	0.17
Middle Archaic Stemmed	−1.09	0.16	0.23	0.14	−0.15	0.12
Late Archaic Stemmed	0.84	0.09	−0.57	0.07	−0.60	0.07
Late Archaic Barbed	−0.30	0.18	−0.75	0.15	0.29	0.14

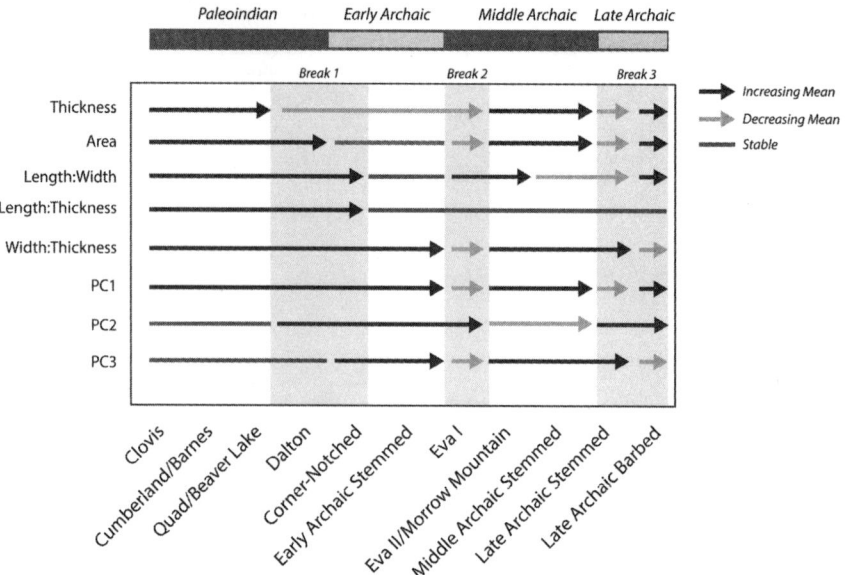

FIGURE 4.17. Trends in the size and shape of the biface groups with three primary breaks indicated.

Middle Archaic. Finally, the values again decrease with the Late Archaic Stemmed and Late Archaic Barbed groups.

For the third principal component, the variation is primarily due to the width:thickness variable. From the Paleoindian through the Early Archaic groups, there is a slight decrease in the relative widths, although this is not statistically significant aside from the two Early Archaic groups being significantly smaller than their Paleoindian counterparts. However, compared to the preceding point types, the relative width of the Eva I group is much larger. Then, following Eva I, the values decrease for the Eva/Morrow Mountain, Middle Archaic Stemmed, and Late Archaic Stemmed types. Finally, with the Late Archaic Barbed groups, the values increase again, indicating an increase in the relative width.

Overall, there appeared to be three major breaks in the size and form of discarded bifaces in the study (Figure 4.17). The first break occurs at the Pleistocene/Holocene boundary. Over the course of the Paleoindian period, area, thickness, and length:thickness exhibit decreasing means and standard deviations, while the length:width is stable. Then, with the

appearance of Dalton and Corner-Notched points that span the Pleisto-
cene/Holocene boundary, the trends reverse, and during the Early Holocene, discarded points are smaller, thicker, and display less length relative
to the width of the points. These variables indicate that over the course
of the Pleistocene/Holocene transition, points show greater evidence
of being discarded at smaller, more resharpened states despite changes
in artifact design. This trend is also reflected in the results of the PCA,
where points display decreasing means and standard deviations over
time in the first principle component (e.g., size) and the second principle
component (e.g., shape) indicates proportionality more consistent with
increasing resharpening over this time.

The second major break occurs during the mid-Holocene, at the
boundary between the Early and Middle Archaic periods. With the Eva I
type, points become thicker with more surface area, and most of that
area comes in the form of increasing width relative to thickness. Surprisingly, the length:width upon discard is indicative of relatively high rates
of resharpening. This is further reflected in the results of the PCA, where
the first principle component increases significantly with Eva I, but the
second principal component shows increasing proportionality indicative of increasing resharpening that begins with Clovis and reaches an
apex with the Eva I group. The third principal component shows larger
increases in relative width, which shows that these points are larger and
much wider than the preceding types.

The third, and most complicated, break occurs with the last two
groups in the study sample—Late Archaic Stemmed and Late Archaic
Barbed. With the Late Archaic Stemmed sample, the area and thickness
variables increase significantly relative to the preceding Eva/Morrow
Mountain and Middle Archaic Stemmed groups. Most of this added
volume comes from the length relative to the width. This is immediately
followed by a decrease in thickness, area, and the length:width with the
Late Archaic Barbed group. In the PCA, the first principal component
(i.e., size) increases with the Late Archaic Stemmed group and then decreases with the Late Archaic Barbed group. However, with the second
principal component, there was a decrease in values that is indicative
of a reduction of relative length with the Late Archaic Stemmed and
Late Archaic Barbed relative to the Middle Archaic Stemmed group. On

the other hand, the third principle component shows that Late Archaic Barbed points are much wider relative to their thickness than the Late Archaic Stemmed points.

Over the course of the Middle and Late Archaic there was 1) a decrease in the mean size of the biface groups and 2) a decrease in the relative amount of length lost to resharpening. This indicates a decrease in the overall amount of unexpended utility in the projectile points that were discarded. Point size then increases with Late Archaic Stemmed and decreases with the Late Archaic Barbed, while the relative length stays somewhat constant. As a result, there is an increase in the amount of unexpended utility with the Late Archaic Stemmed group relative to the Middle Archaic Stemmed and Late Archaic Barbed groups.

Discussion

The three major breaks discussed above coincided with what are likely major changes in subsistence over the Paleoindian and Archaic periods in the Mid-South. First, with the exception of Kimmswick (Graham et al. 1981) and Sloth Hole (Hemmings 1998), there is minimal direct evidence for the exploitation of megafauna in eastern North America. More locally, in the Nashville Basin, the association of stone tools with mastodon remains at Coats-Hines (Deter-Wolf et al. 2011; see also Tune 2015), as well as an overlap with the recovery of Clovis bifaces and mastodon remains recorded in the Tennessee Fluted Points Survey (Breitburg and Broster 1994, 1995), provide more evidence that these larger species were exploited, at least in the initial states of the human colonization of this area.

However, Younger Dryas–aged sites such as Dust Cave and a handful of other cave and rockshelter sites in the Mid-South are notable for their high percentage of smaller, harder-to-catch species, including avian and smaller mammal species (Styles and Klippel 1996; Walker 2007). While the actual zooarchaeological record of the Paleoindian period is sparse, the pattern of the discard of bifaces indicates a broader trend with decreasing variability with the two extremes: Clovis, which are discarded on average larger but with a higher degree of variability, and Dalton, which are smaller, less variable, and exhibit a length:width proportion indicative of higher rates of resharpening.

Consequently, the general pattern of discard reflects artifacts that are likely maintained and reused for longer periods of time over the course of the Paleoindian period, which also coincides with the duration of the Younger Dryas. This conclusion is consistent with the first hypothesis put forth at the beginning of this chapter, which is that from the Bølling-Allerød to the Younger Drays, hunting yields will diminish as a result of the extinction of megafauna and that should be reflected in the biface inventory by repeated attempts to extend the use-lives of the tools. Or, to use the caricature of the rifle-wielding hunter from Chapter 3, this would be the equivalent of going from a situation where a hunter could fire indiscriminately at abundant game to conserving ammunition as game becomes scarcer.

The next major break occurs at the shift between the Early to Middle Archaic, or at the transition between the Early and Middle Holocene. Compared to the preceding point types in the Early Holocene that displayed evidence for extensive resharpening, Eva I points from the peak of the mid-Holocene were discarded on average with a large amount of unexpended utility. In other words, there was a large amount of usable stone available upon discard, but prehistoric hunters made little effort to salvage them for further use. The conforms to the second prediction presented at the beginning of this chapter, which is that environmental conditions during the mid-Holocene should support an increase in hunting yields and that tools should be discarded with minimal evidence for resharpening.

There are several additional lines of evidence to support the conclusion that during the mid-Holocene, prey (and white-tailed deer in particular) became more abundant, and hunters adjusted their stone tool kits accordingly. The closest and most relevant pollen records at Anderson and Jackson ponds display a noticeable increase in the abundance of oak and hickory pollen over the Early Holocene that peaked during the mid-Holocene (Delcourt 1979; Wilkins et al. 1991). While Dust Cave and a handful of other Late Paleoindian assemblages are also well known for their preservation of diverse faunal remains (Styles and Klippel 1996; Walker 2007), one of the key findings of Walker's (1998) analysis was the high percentage of avian species during the Younger Dryas and Early Holocene relative to the components from later time periods.

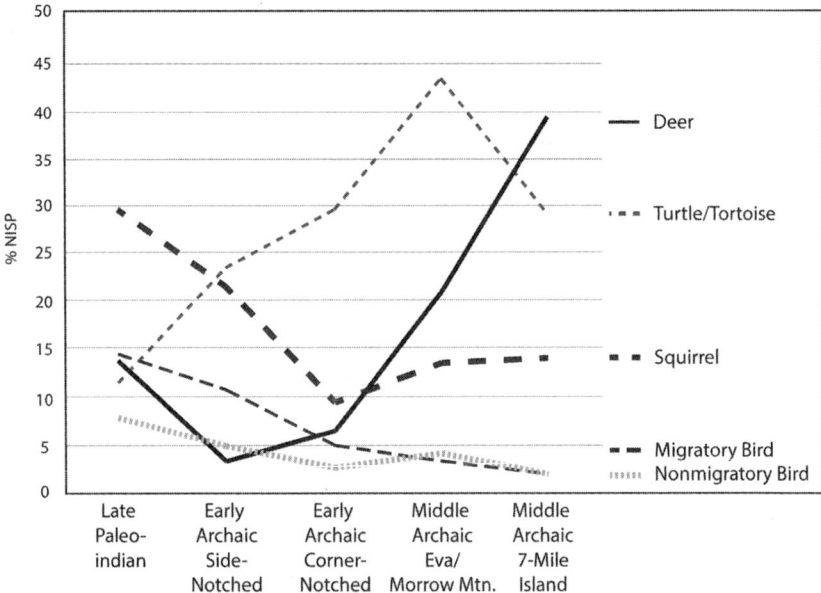

FIGURE 4.18. The percent NISP for the deer, bird, squirrel, and turtle/tortoise categories by component from Dust Cave, Alabama (from Walker 1998).

Following Stiner (2001), I divided Walker's taxonomic classifications into categories based on both size and search and handling costs (Figure 4.18). From the Late Paleoindian through Middle Archaic, there was an increase in the proportion of deer remains relative to smaller, relatively harder to catch species like birds and squirrels. Moreover, the frequency of turtle and tortoise remains also increases over this time period. In the eastern Mediterranean, Stiner argued that turtles and tortoises are highly susceptible to population crashes as a result of human predation because they are relatively easy to catch and have slow reproductive rates. In the case of Dust Cave, species like turtles and tortoises (which are relatively easy to catch), and deer (the largest herbivore) increase in frequency through the Early and Middle Archaic. In other words, the inhabitants of Dust Cave appear to be focusing their efforts on the highest-ranked species in terms of overall caloric return after adjusting for search and handling costs. This pattern is consistent with increasing yields from hunting over time.

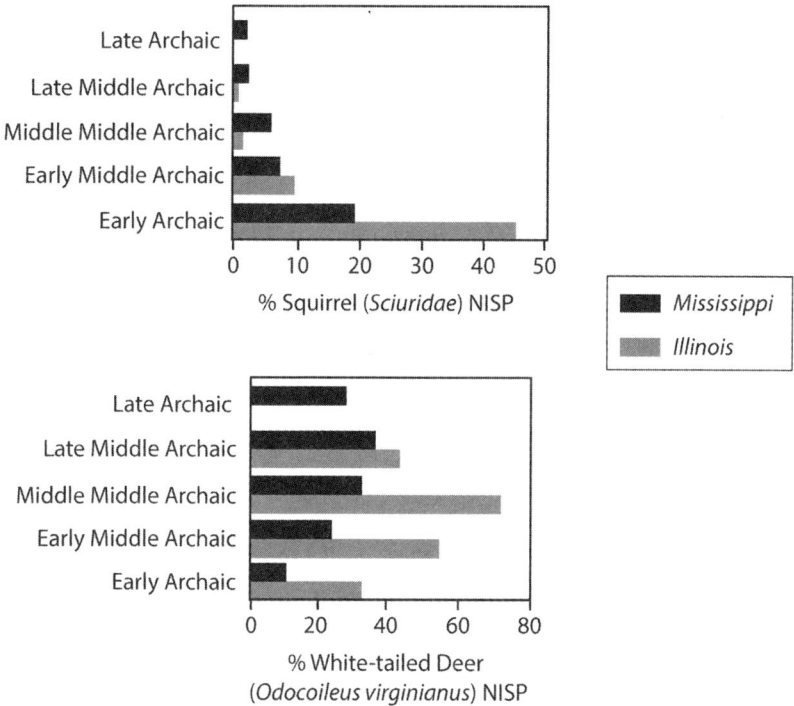

FIGURE 4.19. The percent NISP for squirrel (*top*) and white-tailed deer (*bottom*) categories from a sample of archaeological assemblages in the Illinois and Mississippi River drainages (modified from Styles and Klippel 1996).

The focus on high-ranked fauna is not specific to Dust Cave. Styles and Klippel (1996) examined nine faunal assemblages from across the Mid-South and found a similar trend—most notably, that the relative frequency of deer dramatically increases in the Middle Archaic but then diminishes in later time periods when compared to smaller species, like squirrels (Figure 4.19). Similarly, Bissett (2010) compiled data on a set of Middle Archaic sites from Tennessee, Kentucky, Illinois, and Indiana and found that while the proportion of deer remains clearly dominated these assemblages, by the end of the Middle Archaic, the relative pro-portion of deer remains had already begun to decrease. These broad pat-terns in faunal exploitation are also consistent with the third prediction put forth at the beginning of this chapter, which is that after the peak in the mid-Holocene, yields would be expected to diminish as climate becomes cooler and wetter, which would undermine the frequency and abundance of oak and hickory mast.

Initially, it appears that the biface assemblages from Benton and Humphreys counties follow suit by becoming more conservative with their biface inventories. Following Eva I, the subsequent point groups (Eva/Morrow Mountain and Middle Archaic Stemmed) decrease in both area and thickness, which is also reflected in the results of the PCA, where the first principle component also indicates a reduction in size. There was also a decrease in the width relative to the length and thickness. As a result, the points become relatively narrower but with less overall size. This presents an interesting situation where on one hand, the size of the bifaces conforms to the expectations of the model: they become much smaller over time. On the other hand, the shape of the bifaces do not conform to the expectations of the model; rather than displaying increasing evidence for resharpening, they appear to be discarded more readily.

Over time, hunters stopped stocking their spear points with potential utility (e.g., mass), and they no longer continued to maintain their artifacts after using them. Again, following Kuhn and Miller's (2015) model, this type of pattern would be indicative of a decrease in the expected returns from hunting over the course of the Middle Archaic, but with an importation distinction. When anticipated returns diminished after the mid-Holocene, it appears that the hunters of the lower Tennessee River valley did not consider it a worthwhile investment in time or energy to carry around a lot of excess lithic raw material at the end of their spears when compared to the preceding Eva I group.

To once again return to the analogy of the rife-wielding hunter, the behavior would be the equivalent of realizing that the odds of a successful hunting trip are not good and scaling back on the amount of ammunition to purchase as a result. This is an important distinction, because when faced with diminishing yields in the Younger Dryas and Early Holocene, the hunters of the lower Tennessee River valley sought to extend the use-lives of their artifacts through resharpening. After the mid-Holocene, when hunters were faced with the same problem (diminished yields), they chose an entirely different strategy for their biface inventory: they scaled back the energy they would have directed toward making and carrying large bifaces and presumably directed that energy elsewhere.

This shift in biface technological organization is even more striking given that it coincides with the Middle Archaic Stemmed group, which includes the Benton type (Justice 1995:111; Lewis and Lewis 1961:34). Bentons are known for their relatively large size and exceptional craftsmanship and are often found a significant distance from the sources of raw material from which they were made (McNutt 2008; Meeks 2000). While a number of Benton bifaces were recorded in the study sample, I excluded all points that were found in burial contexts, including several of the type of specimens first identified by Lewis and Lewis (1961:34) from the Eva site. Consequently, the difference in size between bifaces found in mortuary and nonmortuary contexts may have an underlying economic rationale for why Benton bifaces became central to emerging trade networks in the Mid-South (e.g., Jefferies 1996; McNutt 2008). In other words, the production of large, expertly crafted bifaces became more rare, more costly, and more out of the ordinary when compared to the larger trends in biface production, maintenance, and discard during the Middle Archaic.

The third and final break in this dataset occurs with the appearance of Late Archaic Stemmed and Late Archaic Barbed types, where the mean values in the first component (e.g., size) increases significantly with the Late Archaic Stemmed sample and decreases with the Late Archaic Barbed sample. As a result, there appears to be a large increase, and then decrease, in the size of bifaces at the end of the Archaic period. However, in addition to being smaller in size than the Late Archaic Stemmed group, Late Archaic Barbed bifaces are also relatively short and wide. This shift from Late Archaic Stemmed to Late Archaic Barbed reflects an increase and subsequent decrease in the amount of unexpended utility upon discard. Compared to the two previous breaks in biface technological organization, this third break is unexpected when compared to the predictions laid out in the beginning of this chapter. It does not appear to correlate to any obvious environmental trends that are evident in regional proxy records. Moreover, Styles and Klippel (1996) found no change in the relative population of deer in the Late Archaic. Yet the inference from the increase and decrease in the size of the bifaces could potentially indicate variability in hunting yields that are responding to environmental changes that are much subtler when compared to our proxy records, or they are no longer making and using bifaces in response to hunting yields, which would require a new alternative hypothesis to explain their behavior. It is worth noting that the appearance of the large Late Archaic Stemmed bifaces coincides with the dates for the earliest domesticated plants in eastern North America (Smith and Yarnell 2009:6564).

Conclusion

The variation in size and shape of bifaces over the Paleoindian and Archaic periods presents an opportunity to examine trends in hunting yields over time encompassing the first appearance of people in eastern North America and the appearance of domesticated plants. This is made possible by the unique location of the study area, which is located in a region that contains abundant and widely available sources of lithic raw material. By utilizing a model that treats artifacts as potential patches of utility and paleoenvironmental reconstructions to create a null model for how people should organize their biface technology, it is possible to evaluate the choices made about the production, use, maintenance, and discard of these artifacts.

Moreover, there are three major breaks in biface technological organization, the first two of which conform to environmental changes; the third does not. The first occurred during the Younger Dryas after megafauna went extinct. The techniques employed by those who produced and resharpened Late Paleoindian Dalton points and the inception of notching allows for extensions of utility relative to the area and thickness of the point. In other words, this provided a way to extend the use-lives of these bifaces during the Younger Dryas and Early Holocene.

However, with the transition from the Early to Middle Holocene, a second break appears and coincides with points that were discarded with a relatively high degree of unexpended utility. The timing of the break is coeval with a peak in temperate climate in the mid-Holocene, which may have increased the frequency of acorn and hickory masting, resulting in a boom in the availability of deer. The trend then reverses over the remainder of the Middle Archaic, where bifaces are on average smaller, with less evidence of resharpening. This is probably related to the decrease in the relative proportion of deer in faunal assemblages over the Middle Archaic period. Finally, the third break appears in the Late

Archaic, where there is an increase, and then a decrease, in the size of bifaces that does not appear to correlate with changes in the environment observable in regional proxy records for the environment. It does, however, coincide with the appearance of domesticated plants in eastern North America.

Smith (2011) argued that there is minimal evidence for resource stress and population packing in the time periods coinciding with the appearance of domesticated plants during the Late Archaic period. However, the results of this study, combined with zooarchaeological and paleoenvironmental information, show that, over the course of the Paleoindian and Archaic periods, there are broad trends in the design and discard of bifaces and the content of faunal assemblages that are indicative of hunting yields that, for the most part, wax and wane with the availability of prey.

While Smith and Yarnell (2009) and Smith (2011, 2015) have argued that there is no evidence for resource stress in the time periods preceding the appearance of domesticated plants in eastern North America, it is clear that the availability of prey was not constant and that the overall trend from the Middle to Late Archaic was one of diminishing yields. On the other hand, the appearance of domesticated plants occurs during a break in biface technological organization that would appear to point to an increase in hunting yields that is independent of observable changes in the environment. I would argue that stone tool, zooarchaeological, and paleoenvironmental records might indicate an alternative hypothesis: resource availability covaried with the environment to such a degree that it impacted biface technological organization *until the appearance of domesticated plants*. Thus far, the focus of this study has been on resource availability. Now it is time to turn to the other variable that is key to the debate over the origins of domesticated plants in eastern North America: population pressure.

Notes

1. Copies of all of the raw data from this study are curated by the McClung Museum at the University of Tennessee in Knoxville and the Tennessee Division of Archaeology in Nashville.

CHAPTER 5

The Ideal Free Distribution
and Landscape Use in the Duck and
Lower Tennessee River Valleys

Introduction

One key variable in exploring the origins of agriculture across the world is the role of population pressure. V. Gordon Childe (1936) argued for increasing population near rivers and oases as part of his oasis hypothesis, and, later, Binford (1968), Flannery (1969), Rosenberg (1998), and Stiner (2001) argued for population pressure as a catalyst for the broad spectrum revolution that preceded the appearance of domesticated plants and animals. However, there are multiple critiques of this Malthusian explanation, most prominently by Bruce D. Smith (2011, 2015) and Melinda Zeder (2012).

In its most basic formulation, population pressure can be thought of as the amount of resources relative to the individuals who require those resources (e.g., Malthus 1798). In the previous chapter, I focused primarily on the role of resource scarcity, with an emphasis on hunting yields through time and implications for demographic trends in the backdrop. In this chapter, I focus more directly on demographic trends using an alternative source of data: the distribution of archaeological sites across space and time.

In order to generate hypotheses about population growth that can be tested with archaeological data, I rely on a formal model from human behavioral ecology, ideal free distribution (IFD) (e.g., Fretwell and Lucas 1969; Sutherland 1996). The archaeological data in question consists of recorded archaeological sites in the Duck River drainage, which is where the earliest known domesticated sunflower seed was identified at the Hayes site (Crites 1993)

Archaeological Approaches to Demography in Eastern North America

Population pressure is a linchpin in many of the models proposed for the origins of agriculture. Presumably, an increase in population density should be evident in the archaeological record, but several of the most common means for measuring demographic trends are problematic in eastern North America. For example, the frequency of ^{14}C dates is rapidly becoming a common method for extracting demographic trends (Kelly et al. 2013; Surovell et al. 2009). In some instances—most notably, Australia (Williams 2012, 2013) and the Neolithic expansion into Europe (Shennan and Edinborough 2007)—this approach has been used effectively. However, taphonomic bias (Surovell and Brantingham 2007; Surovell et al. 2009) and differential sampling (Ballenger and Mabry 2011) must be taken into account before assuming that the frequency of radiocarbon dates translates directly to the number of people at a given point in time.

In eastern North America, this is particularly difficult—especially for the Paleoindian and Early Archaic period—because sample sizes are small and the effects of sampling and taphonomic biases likely swamp any demographic signal in some regions (Miller and Gingerich 2013b). However, despite the difficulty in accounting for these factors, they can be overcome. As an example, Weitzel and Codding (2016) used a database of radiocarbon dates and archaeological sites to argue that the 1,000 years before the origins of agriculture likely saw an increase in population.

Another potential source of information is bioarchaeological data. Maria O. Smith (1996:134) states that the available records in eastern North America are "often spotty and focused on large cemetery samples." While many of the large cemetery samples are located in the Mid-South, including Indian Knoll in Kentucky and Eva in Tennessee, Smith contends that the "most anyone could say about Archaic populations was that they were robust and had high levels of dental attrition, few caries, and long heads." Some recent studies, however, focus on demography and population structure using much larger and geographically expansive datasets. For example, Joseph F. Powell (1995) used a large sample of sites from across eastern North America and found that during the Middle and Late Holocene, gene flow likely increased as populations became more fixed on the landscape. On the other hand, Nicholas P. Herrmann (2002) found significant differences between the Green River sites in Kentucky and the Eva site. He argues that within the Green River sample, the population is much more homogenous and likely indicates local mate exchange networks and greater female mobility. However, he states that before biological data can be used to make more definitive statements about population structure and demographic trends, more existing collections need to be reanalyzed and a much more extensive dataset (distributed both through time and across space) is necessary.

Another strategy for examining demographic trends in eastern North America has been to examine the frequency and distribution of archaeological sites. For example, Anderson (1996) examined the site file records from states in the Lower Southeast and was able to generate a database of 32,428 Archaic period sites, ranging from isolated artifacts to dense middens. When standardized by time period, he found that sites decrease in frequency from the Early to Middle Archaic but increase more than twofold in the Late Archaic period. Moreover, the Early Archaic sample is widely but unevenly distributed over the landscape and has a tendency to occur along major rivers, with some concentrations at or near known major lithic raw material sources. Middle Archaic sites in the Mid-South, on the other hand, appear to be much more restricted and localized, and concentrations of sites rarely extend more than a few counties, especially in the Tennessee, Cumberland, Duck, and Green River drainages. These locations overlap with major known areas where shell- and earth-midden deposits occur in relatively high frequencies. Finally, Late Archaic sites occur widely, which led Anderson (1996:165) to argue that, by this time, "moderate to extensive use of almost every part of the region is indicated, suggesting that considerable landscape filling had occurred."

While using the frequency and distribution of archaeological sites and artifacts to assess demography is useful because of their much larger sample sizes and distributions relative to other proxies, there are still several issues that must be overcome with this approach. First, like the frequency of ^{14}C dates, archaeological sites are also subject to a variety of taphonomic biases that make finding and recording older archaeological sites more difficult (e.g., Dunnell 1990; Kelly et al. 2013; Surovell et al. 2009). Moreover, as Anderson (1996) frequently mentions in his analysis, certain areas have been subjected to substantially more research than others, which leads to a form of survey and research bias in the sample. Another issue noted by Anderson is the variation in the size of sites (isolated finds versus large, dense sites) and the fact that state site files usually only record temporal data by broad time periods.

One way to resolve the differential reporting of archaeological sites would be to convert a site-based approach into a non-site survey (e.g., Dunnell and Dancey 1983; Thomas 1975). For example, Cabak and colleagues (1996) analyzed the site records for the Savannah River site to make inferences about shifts in landscape use over the duration of the Holocene. Another potential analytical avenue would be to use Surovell's (2009) proxies for measuring occupation intensity for a sample of assemblages spanning the Holocene. However, attempting to replicate Cabak and colleagues' (1996) and Surovell's (2009) analyses—particularly in the Mid-South—would be difficult because of different excavation protocols and curation criteria would severely limit the number of sites available for analysis.

Finally, attempts to use the distribution and frequency of sites to reconstruct demography suffer from the lack of a developed interpretive framework. For example, Meltzer (1988) interpreted the high frequency of recorded Early Paleoindian sites and their wide distribution across the southeastern United States as evidence that groups were smaller and more residentially mobile when compared to Paleoindian groups at higher latitudes. Conversely, Anderson (1996) interpreted the increase in the frequency and wider distribution of sites in the Late Archaic across the Southeast as evidence for an increase in population compared to earlier time periods. Does the increase in frequency and a wider distribution mean more people or fewer people moving around more often? Clearly, there is an equifinality issue here. To solve this problem, the ideal free distribution (Fretwell and Lucas 1969) can be appropriated as a null model to analyze variation in site frequencies and distributions to understand fluctuations in demography.

The Ideal Free Distribution

The ideal free distribution was first used to model habitat selection by birds (Fretwell and Lucas 1969; Figure 5.1). After calculating the relative suitability of the different habitats using factors that include access to resources and potential hazards, the model assumes that people should first settle in areas with the highest suitability. Then, as population increases, resource competition and interference depress suitability. When the suitability of the primary habitat is equal to the next suitable habitat, individuals should then expand into the next suitable habitat.

From the IFD model, there are five predictions for the dispersion of populations into new habitats. First, the highest-ranked habitat should be occupied first and will be continuously occupied. Second, lower-ranked habitats should be inhabited in the order of their relative suitability. Third, population density should be highest in the highest-ranked habitat. Fourth, suitability should equalize among all occupied habitats. Fifth, suitability declines across all occupied habitats and population grows.

In addition to the IFD, there are three other variants of the model. The ideal despotic distribution predicts how habitats should be occupied if one or more individuals are able to control a disproportionate amount of resources. The IFD with an Allee effect models the effect of economies of scale, where increasing population actually increases the suitability of a habitat to a point, followed by diminished suitability when the addition of new individuals adds no additional benefit. Finally, the IFD can be applied to model the effect of innovations that change the suitability of marginal habitats or increase the suitability of occupied habitats.

The IFD's most basic assumption is that individuals will select habitats to maximize fitness and that the suitability of the habitat and population density will influence an individual's decision to either stay in a habitat or move to a location with greater net fitness benefits. The IFD model shares certain characteristics with the marginal value theorem (e.g., Charnov 1976; Kelly 1995:90–97): it assumes that individuals have all of the available information to make a decision on whether to move or stay and that all individuals are free to leave or enter a new habitat. However, unlike the marginal value theorem, the ideal free distribution incorporates expectations relating to population density, or landscape infilling, which makes it useful as an interpretive framework for assessing demography.

The applicability of using the IFD in conjunction with archaeological data has been demonstrated with recent studies of the transition to agriculture in Spain (McClure et al. 2006), the spread of agriculture into Europe (Shennan and Edinborough 2007), the colonization of Oceania (Kennett et al. 2006), the settlement of the Channel Islands in California (Jazwa et al. 2016; Kennett and Winterhalder 2008; Winterhalder et al. 2010), the spread of people into California in general (Codding and Jones 2013), the introduction of horses and snowmobiles as new hunting technologies in North America (Ross and Winterhalder 2015), prehistoric Australia (O'Connell and Allen 2012), the Classic period Maya (Prufer et al. 2107), the Alaskan Arctic Small Tool tradition (Tremayne and Winterhalder 2017), and, more broadly, the demographic consequences of food storage in pre-agrarian societies (Winterhalder et al. 2015).

The Duck River drainage is a tributary of the Tennessee River in central Tennessee and represents an ideal location to examine trends in the frequency and spatial distribution of archaeological sites in the time periods leading up to the emergence of domesticated plants in eastern North America (Figure 5.2). First, the earliest domesticated sunflower remains currently known were recovered at the Hayes site in this drainage. Second, the drainage also intersects the two counties (Benton and Humphreys) from which I drew the sample of sites that I used in the previous chapter to analyze trends in biface technology for the Paleoindian and Archaic periods.

A critical component of using the IFD to interpret archaeological site distributions is determining the suitability of each habitat. Complicating matters further, this study uses assemblages that span approximately 10,000 years, and in Chapter 2, I discussed the substantial environmental changes that occurred over this span of time in eastern North America (e.g., Delcourt and Delcourt 1983, 1985; Viau et al. 2006; Williams et al. 2004). As a function of increasing temperature and moisture, the boreal forests that covered eastern North America during the Last Glacial Maximum began moving north and more temperate species expanded to cover large swaths of land by the mid-Holocene.

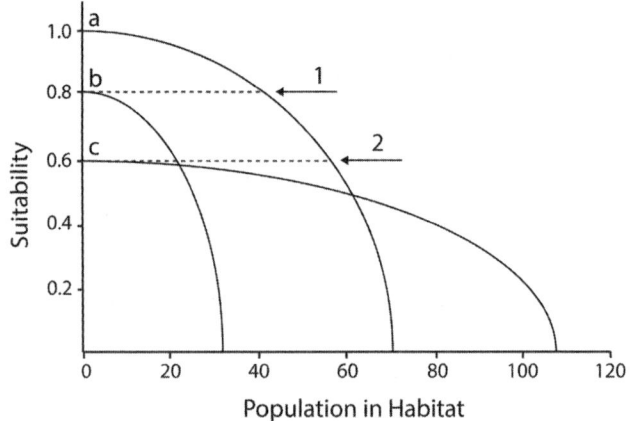

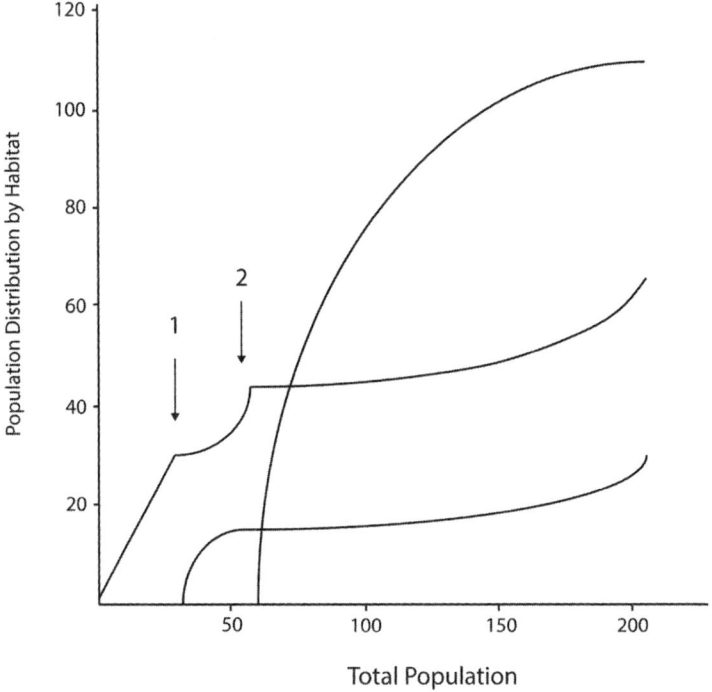

FIGURE 5.1. The ideal free distribution. The upper panel shows suitability curves on a normalized scale of 0–1 for three habitats (a, b, and c) as a function of population density in the habitat. The lower panel shows how population growth will be allocated among habitats given the suitability (Fretwell and Lucas 1969; Winterhalder et al. 2010:473).

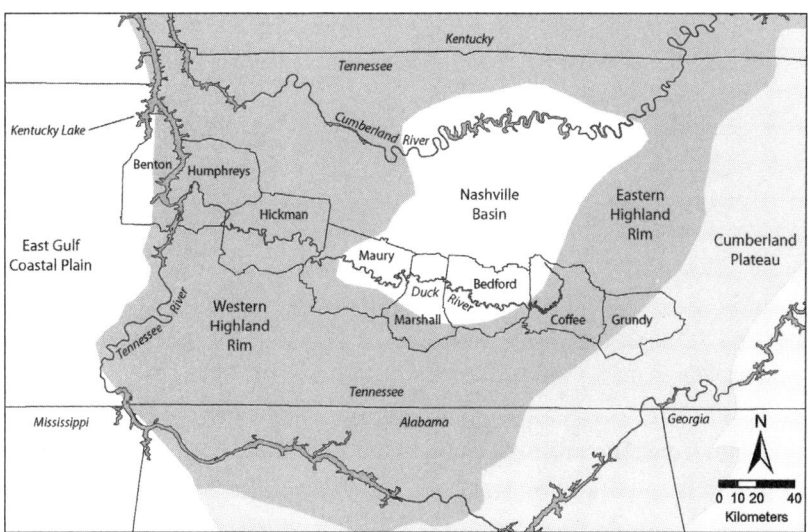

FIGURE 5.2. The Duck River drainage study area.

The movement of boreal and temperate species is also reflected in the distribution of species relative to altitude. For example, Mills and Delcourt (1991) observed that several areas of the Blue Ridge Mountains show evidence of alpine tundra being replaced by boreal forests after 12,500 years ago (~14,761 cal BP). Moreover, the reconstructions by Delcourt and Delcourt (1983, 1985) and Williams and colleagues (2004) show a delay in the northward movement of boreal forests in areas with higher altitude, especially the Appalachian Highlands. The results of these studies are suggestive of a sky island effect, where remnant boreal forests and tundra are present in the uplands well after deciduous forests spread into the lower elevations. However, with the dramatic increase in temperature at the end of the Younger Dryas and into the Early Holocene, the size of these boreal sky islands would have diminished as deciduous forests climbed to higher elevations. A sky-island effect is relevant to the current study because there is a considerable variation in elevation from the eastern Gulf Coastal Plain (~115 m) to the Cumberland Plateau (~600 m) (Figure 5.3).

Multiple studies have also found evidence for fluctuations in environmental conditions during the mid-Holocene (Brackenridge 1984; Delcourt 1979; Klippel and Parmalee 1982). During the peak of the

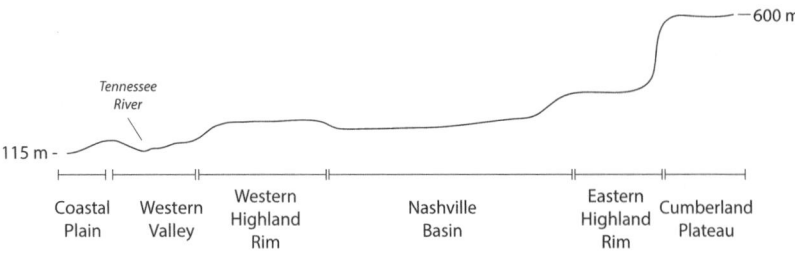

FIGURE 5.3. The variation in elevation across the Duck River drainage (vertical exaggeration 10×).

mid-Holocene, the warming trend that began in the Early Holocene continued, but conditions also became much drier. In the Nashville Basin, this likely caused a retraction of oak-hickory-dominated forests and an expansion of cedar glades. At the end of the mid-Holocene, climate became cooler and moister, causing an expansion of oak-hickory forests in the Nashville Basin.

Based on the environmental variability from the Late Pleistocene to Late Holocene epochs, the IFD can be used to create hypotheses for how people should have expanded into the lower Tennessee and Duck River drainages. First, during Bøllig-Allerød and the Younger Dryas, the appearance of mixed deciduous forests at lower elevations and boreal forests at higher elevations would have made the lower elevations much more suitable for hunter-gatherers. As a result, individuals should have expanded to the lowest elevations first.

Second, as boreal forests at higher elevations are replaced by deciduous forests during the Younger Dryas and Early Holocene, the IFD would predict that individuals should begin expanding to higher elevations once their suitability equals that of the lower elevation areas. Moreover, individuals would not be expected to abandon lower elevation habitats in favor of habitats at higher elevations.

Third, during the mid-Holocene, the overall suitability of the Nashville Basin decreased as primary food sources (acorns and hickory nuts) for both humans and deer became much less common in this physiographic province. Consequently, the IFD model would predict that a drop in suitability in the Nashville Basin would lead to the abandonment of this habitat *if* the population in surrounding habitats was low enough to absorb new occupants without affecting the suitability of these habitats.

Finally, after the peak of the mid-Holocene, and with the retraction of the cedar glades in the Nashville Basin in favor of oak and hickory forests, the environmental suitability of the Nashville Basin should increase. Consequently, individuals should be expected to be present in the Nashville Basin once again as oak- and hickory-dominated forests re-expand into these areas.

Study Sample

Works Progress Administration crews first surveyed the areas around the confluence of the Duck and Tennessee rivers in the 1930s as part of the Kentucky Lake project (Lewis and Kneberg 1959). It has been resurveyed by Cultural Resource Analysts Inc. in the early 1990s (Kerr and Bradbury 1998) and most recently by the University of Tennessee Archaeological Research Lab (Angst et al. 2011). The section of the river in Coffee and Bedford counties in the Eastern Highland Rim was surveyed as part of the Normandy Reservoir project in the early 1970s (Faulkner and Mc-Cullough 1973). The section of the drainage that traverses the Nashville Basin in Maury and Marshall counties was surveyed as part of the Columbia Reservoir project (Klippel and Parmalee 1982). Robert L. Jolley (1980) conducted a survey of sites in Hickman and Humphreys counties to locate sites between the boundaries of the Kentucky Lake and Columbia Reservoir surveys. In the ensuing years, many additional sites were reported as part of cultural resource management projects, making this one of the more comprehensively surveyed drainages in the Mid-South. The Duck River also crosses multiple physiographic provinces, including the lower Tennessee River valley, the Western Highland Rim, the Nashville Basin, the Eastern Highland Rim, and the Cumberland Plateau (e.g., Fenneman 1938). Because of the unique geologic setting and variation in elevation in this drainage, each physiographic section can be used as a proxy for a habitat in an ideal free distribution analysis.

There are two primary sources of data on the distribution of sites and artifacts for the study area. The first is the Tennessee state archaeological site files at the Tennessee Division of Archaeology. It contains information on site location, condition, cultural affiliation, artifacts recovered or

TABLE 5.1. Summary data for the site file record analysis for the Duck River drainage.

Temporal Component [1]	Coastal Plain/Western Valley	Western Highland Rim	Nashville Basin (Maury County)	Nashville Basin (Marshall and Bedford Counties)	Eastern Highland Rim (Coffee County)	Eastern Highland Rim & Cumberland Plateau (Grundy County)	Total
Clovis/Gainey/Redstone	43	5	0	1	4	0	53
Cumberland/Barnes	26	2	2	0	5	0	35
Quad/Beaver Lake	31	9	7	1	12	0	60
Dalton/Greenbrier	33	7	1	2	13	2	58
Early Archaic Corner-Notched	26	22	23	29	61	13	174
Bifurcate	4	4	6	3	18	6	41
Kirk Stemmed & Serrated/Stanly Stemmed	18	14	7	11	16	3	69
Eva I/Eva II/Morrow Mountain	29	14	18	15	30	12	118
Middle Archaic Stemmed	24	49	28	21	46	12	180
Late Archaic Stemmed/Barbed	75	107	52	42	84	27	387
Total Paleo and Archaic	309	233	144	125	289	75	1175
Other	213	166	113	118	185	73	868
Unknown	257	170	261	209	101	29	1027
Total components	1088	802	662	577	864	252	4245
Total sites	568	411	434	373	316	109	2211

[1] A component is determined by the presence or absence of a temporally diagnostic artifact at a site as defined by the Tennessee Division of Archaeology.

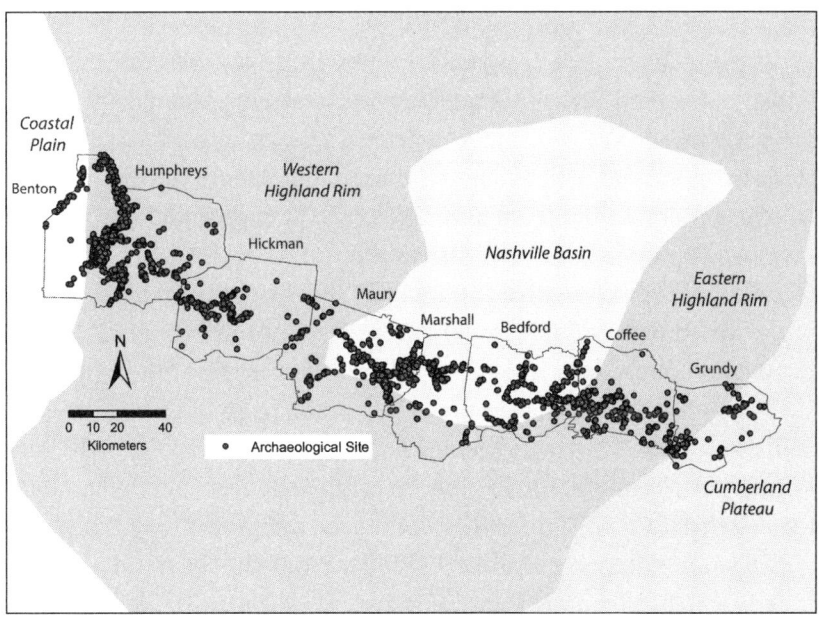

FIGURE 5.4. The distribution of recorded archaeological sites in the Duck River drainage.

observed, and where the collections are curated. Many of these records are in hard-copy format, so in August 2010, I analyzed these files and generated a coding sheet in Microsoft Access that I used to record information on cultural affiliation, the presence of temporally diagnostic artifacts, the presence or absence of shell deposits or earthen mounds, the date of discovery, and when the site file was last updated. In addition to site file information, the state also has one of the most active Paleo-indian projectile point surveys in North America (Anderson et al. 2010; Broster and Norton 1996; Broster et al. 2013). This database, obtained in large part from private collections, contains information on the projectile point type, metric attributes, raw material type, and date of recovery. When applicable, data on Paleoindian points were integrated into the Microsoft Access database to make it comparable to information derived from the state site files. From these sources of information, I created a database of 2,427 localities from the eight counties spanning the extent of the Duck River drainage. Of these, I acquired specific coordinates for 2,211 sites (Figure 5.4; Table 5.1).

Methods

I analyzed several variables that allow me to assess potential sources of sampling bias in the distribution of sites in the study sample. These include temporal and research biases that could influence the frequency in which sites are encountered and reported. These potential effects were assessed by examining 1) the distribution of dates at which sites were recorded, 2) the overall distribution of all sites by physiographic section, 3) the distribution of all sites by component, 4) the frequency of sites relative to county area size, 5) population density from the U.S. Census by county (U.S. Census Bureau 2010), 6) the frequency of sites by land cover (USGS 2013a), 7) geologic formation age (USGS 2013b), and 8) the relative amount of alluvium to uplands in each county based on the USDA-NRCS Soil Survey Geographic Database (SSURGO) (USDA-NRCS 2013). Where appropriate, I used Pearson's chi-square goodness of fit test to assess for statistical significance (Pearson 1900; Plackett 1983) and Cramér's V, which converts the results of the chi-square test into a value that ranges from 0 to 1 (Cramér 1946).

As described in the results section below, there appears to be differential reporting of sites in the Duck River drainage. As a means of counteracting this problem, I first divided each cell by the total number of Paleoindian and Archaic components identified in each county. For example, in the Coastal Plain physiographic section, there are 43 Clovis components out of a total of 309 Paleoindian and Archaic components for the entirety of the Duck River drainage. For that cell, I simply divided the number of Clovis components by the total number of Paleoindian and Archaic components, which provides a relative measure of the distribution of components over time for the Coastal Plain physiographic section.

Alternatively, I divided each cell by the total number of sites for the same temporal components. Again, using the Clovis sample from the Coastal Plain physiographic section as an example, I divided the total number of Clovis components in the physiographic section (43) by the total number of Clovis components in the entire study sample (53). This provides a relative measure of the distribution of components across space at a particular time. Finally, I used Pearson's goodness of fit test

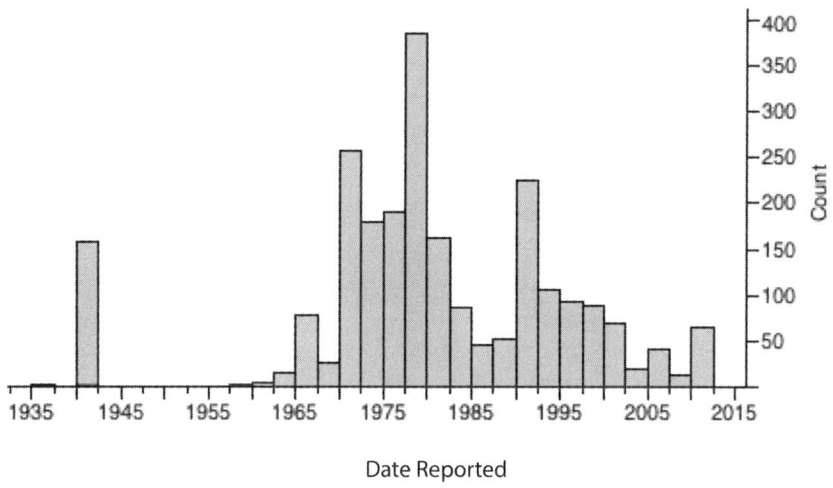

FIGURE 5.5. The frequency of sites in the Duck River drainage by date reported from the Tennessee Division of Archaeology site file records.

(Pearson 1900) with Cramér's V (Cramér 1946) as another way to statistically determine the degree to which sites are evenly distributed across the drainage by temporal component. For this analysis, I compared the reported number of sites against an even distribution of sites across the entire drainage.

Results

Based on this analysis, differential reporting of sites across the Duck River drainage is occurring. For example, there are several peaks and gaps in the reporting of sites over time (Figure 5.5). A peak that occurs in the late 1930s and early 1940s is due to the initial survey of the lower Tennessee River in advance of the creation of the Kentucky Lake reservoir. Following this survey, no sites were reported in the Duck River drainage until the mid-1950s. From the mid-1960s through the early 1980s, the majority of the sites in the Duck River drainage were identified and reported as a result of the Columbia (Klippel and Parmalee 1982) and Normandy (Faulkner and McCullough 1973) reservoir projects and also Jolley's (1980) survey of the lower Duck River. Two additional peaks occur in the early 1990s and 2010, which coincide with additional TVA-sponsored

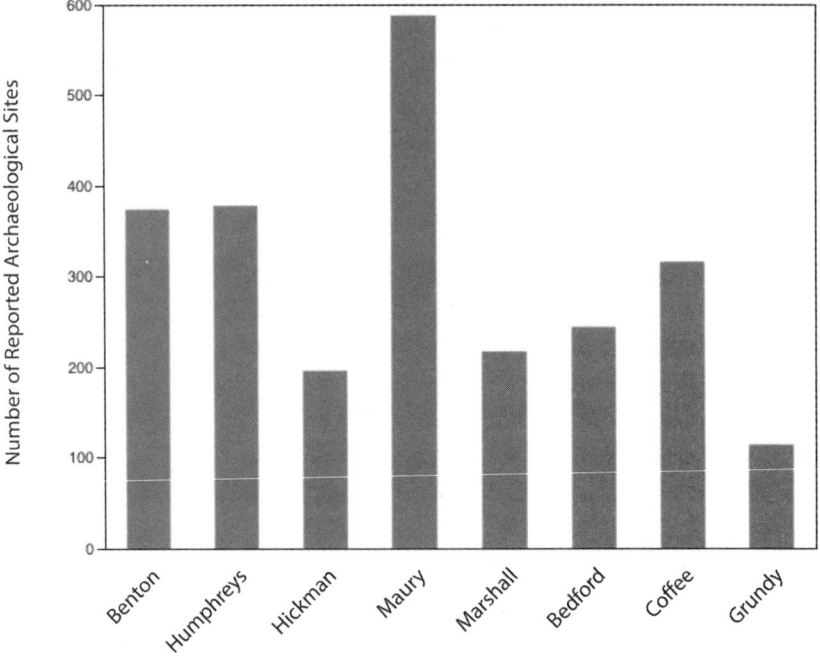

FIGURE 5.6. The frequency of reported sites by county in the Duck River drainage from the Tennessee Division of Archaeology site file records.

surveys of Kentucky Lake by CRA (Kerr 1996) and the Archaeology Research Laboratory at the University of Tennessee (Angst et al. 2011).

As for the distribution of reported sites across the drainage, 26 percent of the sites are located in the Coastal Plain/Western Valley physiographic section (Figure 5.6). This is likely due to the three major surveys of the Kentucky Lake reservoir and active cultural resource monitoring by the TVA. The next highest concentration of sites (19 percent) occurs in the Nashville Basin physiographic section within Maury County, with most of the sites recorded during the Columbia Reservoir project, which also included sites in Marshall County. The Western Highland Rim, the Eastern Highland Rim, and the Cumberland Plateau physiographic provinces are the most underrepresented areas. The variation between physiographic sections is statistically different from a null model based on an even distribution of sites across the drainage. For example, if a particular physiographic section has 30 percent of the total area of the

TABLE 5.2. Chi-square analysis for area size and site frequency in the Duck River drainage.

Physiographic Section	Area (square km)	Area (%)	Sites (observed)	Sites (expected)	Differential	X^2
Coastal Plain/ Western Valley	1511.78	15.08	568	333.50	234.50	164.90
Western Highland Rim	3950.52	39.42	411	871.48	460.48	243.31
Nashville Basin (Maury County)	714.88	7.13	434	157.70	276.30	484.09
Nashville Basin (Marshall & Bedford Counties)	1589.29	15.86	373	350.59	22.41	1.43
Eastern Highland Rim (Coffee County)	1321.47	13.18	316	291.51	24.49	2.06
Eastern Highland Rim and Cumberland Plateau (Grundy County)	934.82	9.33	109	206.22	−97.22	45.83
Total	10022.77	100.00	2211	2211		

Note: X^2: 941.61; *df*: 5; *p*: <0.001; *V*: 0.46

drainage, it should contain 30 percent of the sites. Using this hypothetical distribution as a baseline for comparison, it is clear that sites are not evenly distributed across the drainage (X^2 = 941.61; *df* = 5; *p*<0.0001; *V* = 0.46; Table 5.2). This is due mostly to an overrepresentation of sites in the Coastal Plain/Western Valley and the Nashville Basin in Maury County and an underrepresentation of sites in the Western Highland Rim and Eastern Highland Rim/Cumberland Plateau physiographic sections.

Moreover, the distribution of temporal components found at these sites is not even through time (Figure 5.7). This is more likely due to a temporal taphonomic bias, whereby older sites are generally more difficult to locate because of preservation issues (e.g., Surovell et al. 2009). In other words, there are more Late Archaic (n = 389) than Clovis (n = 54) components because there has been more time for erosion and other processes to destroy Clovis sites. Another source of variation is that the time ranges for each projectile point type are not defined. For example, Clovis may have spanned only a few centuries (e.g., Waters and Stafford 2007), whereas the Late Archaic types span two millennia (Anderson

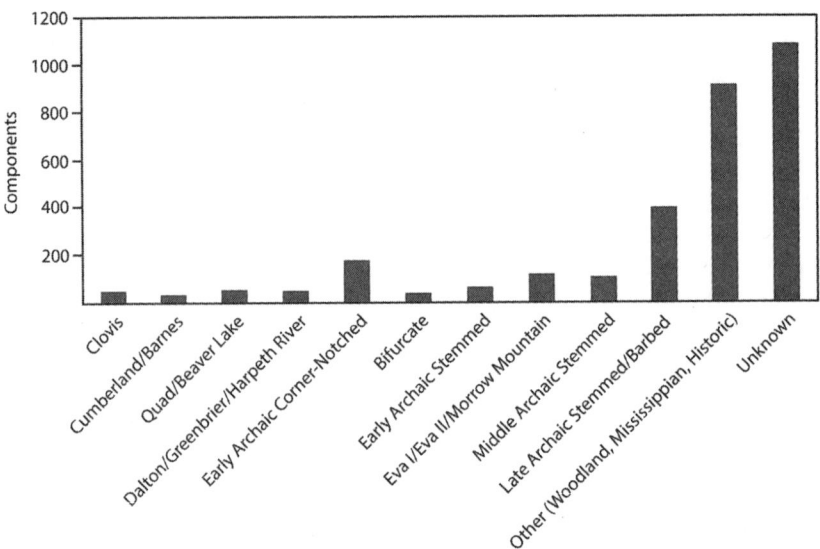

FIGURE 5.7. The frequency of temporal components in the Duck River drainage.

and Sassaman 2012). However, determining an actual range for many of the components in the study sample is problematic because of the limited number of radiocarbon dates in the southeastern United States. This makes it difficult to generate statistically robust age ranges for the temporal components used in the study sample (e.g., Prasciunas 2008; Williams 2012).

Briggs Buchanan (2003), Mary M. Prasciunas (2011), and Michael J. Shott (2004) have all argued that modern population could potentially influence the recovery of Paleoindian period bifaces and, more broadly, the identification of archaeological sites. To test for bias, I examined the distribution of the modern population by county from the 2010 U.S. Census Bureau to the distribution of archaeological sites in the study sample. To construct a Pearson's chi-square goodness of fit test, I created a hypothetical distribution based on the modern population. For example, since Benton County contains 60 percent of the modern population, it should be expected to contain 60 percent of the archaeological sites in the sample if modern population perfectly predicts the identification and reporting of sites (Table 5.3). The results indicate that the distribution of archaeological sites differs significantly from the modern population, a

TABLE 5.3. Chi-square analysis for modern population size and site frequency by county in the Duck River drainage.

County	2010 Population		All Sites			
	n	%	Observed	Expected	Differential	X^2
Benton	16489	5.92	334	130.85	203.15	315.37
Humphreys	17929	6.44	336	142.28	193.72	263.75
Hickman	24690	8.86	180	195.94	−15.94	1.30
Maury	80956	29.06	545	642.46	−97.46	14.78
Marshall	26767	9.61	195	212.42	−17.42	1.43
Bedford	45058	16.17	216	357.57	−141.57	56.05
Coffee	53016	19.03	296	420.73	−124.73	36.98
Grundy	13703	4.92	109	108.75	0.25	0.00
Total	278608	100.00	2211	2211		

Note: X^2: 689.66; *df*: 7; *p*: <0.001; *V*: 0.39

relationship driven for the most part by the high frequency of sites from Benton and Humphreys counties and the comparatively low frequency of sites found in Coffee and Grundy counties. This overrepresentation of sites in Benton and Humphreys counties is likely due to the multiple intensive surveys related to Kentucky Lake on the lower Tennessee River.

This overrepresentation of sites in the lower Tennessee River valley is also evident in the distribution of sites by land cover (Table 5.4). Specifically, while the sites are slightly overrepresented in grasslands and underrepresented in deciduous forests and mixed forests, sites in water bodies are very highly overrepresented. This is likely due to the large number of sites that were initially identified by the WPA in the Kentucky Lake survey and subsequent surveys by CRA and the Archaeological Research Laboratory at the University of Tennessee when lake levels were down. However, these areas would be classified as water bodies now, since the area is either completely, partially, or periodically submerged. Again, since most of the sites in the study sample were recorded in advance of reservoir projects, this is likely a contributing factor as to why sites are overrepresented in alluvium as derived from the SSURGO database (USDA-NRCS 2013; Table 5.5) and in Holocene and Quaternary deposits as opposed to other geologic formations as classified by the USGS (Table 5.6). However, sites are also overrepresented in areas classified as Ordovician-aged, which is likely a result of survey efforts in association with the Columbia Reservoir project in the Nashville Basin.

TABLE 5.4. Chi-square analysis for land cover and site frequency by physiographic section in the Duck River drainage.

| Land Cover Category | Area | | All Sites | | | |
	Km	%	Observed	Expected	Differential	X^2
Urban and built-up land	86.36	0.86	27	19.05	7.95	3.31
Dryland cropland and pasture	104.19	1.04	18	22.99	−4.99	1.08
Cropland/grassland mosaic	17.57	0.18	18	3.88	14.12	51.47
Cropland/woodland mosaic	104.29	1.04	41	23.01	17.99	14.06
Grassland	2.00	0.02	6	0.44	5.56	69.90
Savanna	38.44	0.38	40	8.48	31.52	117.14
Deciduous broadleaf forest	8727.83	87.10	1788	1925.69	−137.69	9.84
Evergreen needleleaf forest	289.70	2.89	67	63.92	3.08	0.15
Mixed forest	442.91	4.42	36	97.72	−61.72	38.99
Water bodies	207.66	2.07	170	45.82	124.18	336.57
Total	10020.96	100.00	2211	2211		

Note: X^2: 642.51; df: 9; p: <0.001; V: 0.38

TABLE 5.5. Chi-square analysis for soils (upland soils versus alluvium) and site frequency by physiographic section in the Duck River drainage.

| Soils Category | Area | | All Sites | | | |
	Km	%	Observed	Expected	Differential	X^2
Alluvium	1637.46	16.34	1455	361.29	1093.71	3310.99
Upland	8383.50	83.66	756	1849.71	−1093.71	646.70
Total	10020.96	100.00	2211	2211.00		

Note: X^2: 3957.69; df: 1; p: <0.001; V: 0.95

TABLE 5.6. Chi-square analysis for geologic formation age and site frequency by physiographic section in the Duck River drainage.

| Geologic Formation Age | Area | | All Sites | | | |
	Km	%	Observed	Expected	Differential	X^2
Holocene	165.87	1.73	252	38.32	213.68	1191.60
Quaternary	441.62	4.61	338	102.02	235.98	545.83
Pennsylvanian	682.93	7.14	35	157.77	−122.77	95.53
Mississippian	4434.65	46.33	404	1024.46	−620.46	375.78
Devonian	132.83	1.39	47	30.69	16.31	8.67
Silurian	99.74	1.04	28	23.04	4.96	1.07
Ordovician	3613.22	37.75	1107	834.70	272.30	88.83
Total	9570.86	100.00	2211	2211.00		

Note: X^2: 2307.32; df: 6; p: 0; V: 0.72

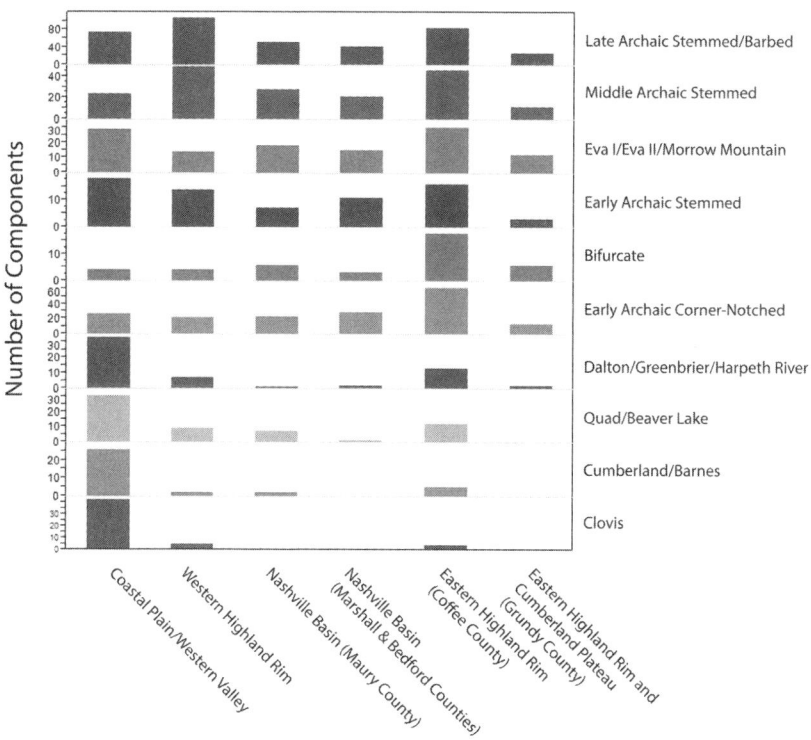

FIGURE 5.8. The distribution of temporal components across each physiographic section.

Based on the analyses above, a temporal taphonomic bias and an overrepresentation of sites in areas that were surveyed as part of the Kentucky Lake, Columbia, and Normandy reservoir projects are the two main sources of variation that appear to be biasing the study sample (Figure 5.8). Consequently, it is necessary to adjust the study sample in two ways to winnow out the effects of these biases in the reporting of sites containing Paleoindian and Archaic components in the Duck River drainage. First, I divided each component in each physiographic section by the total number of Paleoindian and Archaic components in the same physiographic section (Figure 5.9). This accounts for the variation related to survey coverage, area size, modern population density, geology, land cover, and the amount of area that is composed of floodplain alluvium as opposed to more stable, upland surfaces.

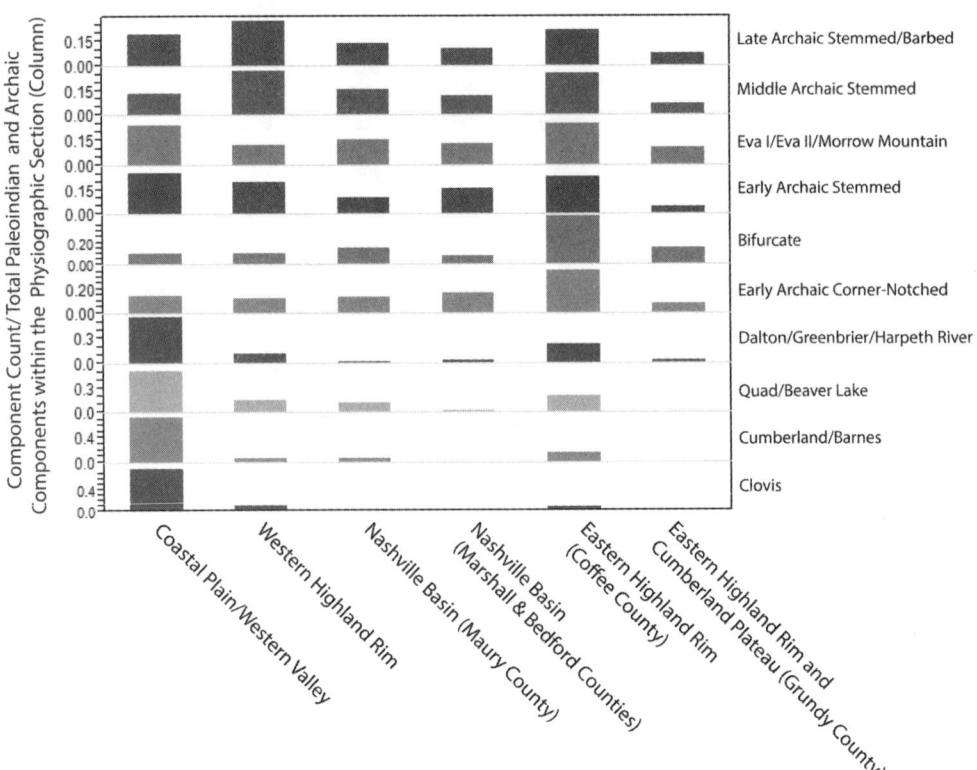

Component Count/Total Paleoindian and Archaic
Components within the Physiographic Section (Column)

Late Archaic Stemmed/Barbed

Middle Archaic Stemmed

Eva I/Eva II/Morrow Mountain

Early Archaic Stemmed

Bifurcate

Early Archaic Corner-Notched

Dalton/Greenbrier/Harpeth River

Quad/Beaver Lake

Cumberland/Barnes

Clovis

Coastal Plain/Western Valley

Western Highland Rim

Nashville Basin (Maury County)

Nashville Basin (Marshall & Bedford Counties)

Eastern Highland Rim (Coffee County)

Eastern Highland Rim and Cumberland Plateau (Grundy County)

FIGURE 5.9. The distribution of temporal components by county divided by the total Paleoindian and Archaic components from within the physiographic section.

Three trends are apparent after adjusting the data to account for temporal taphonomic and survey coverage biases. First, the majority of the Clovis, Cumberland/Barnes, Quad/Beaver Lake, and Dalton/Harpeth River/Greenbrier components are found in the Coastal Plain/Western Valley physiographic section associated with the lower Tennessee River valley. However, there is a slightly higher representation of the Quad/Beaver Lake and Dalton/Harpeth River/Greenbrier components across the drainage, especially in the Eastern Highland Rim in Coffee County. Then, with the Early Archaic Corner-Notched and Bifurcate components, the trend reverses with the majority of the sites in these time periods located in the Nashville Basin, the Eastern Highland Rim, and the Cumberland Plateau. Beginning with the Early Archaic Stemmed group, the subsequent components are more widely distributed across the drain-

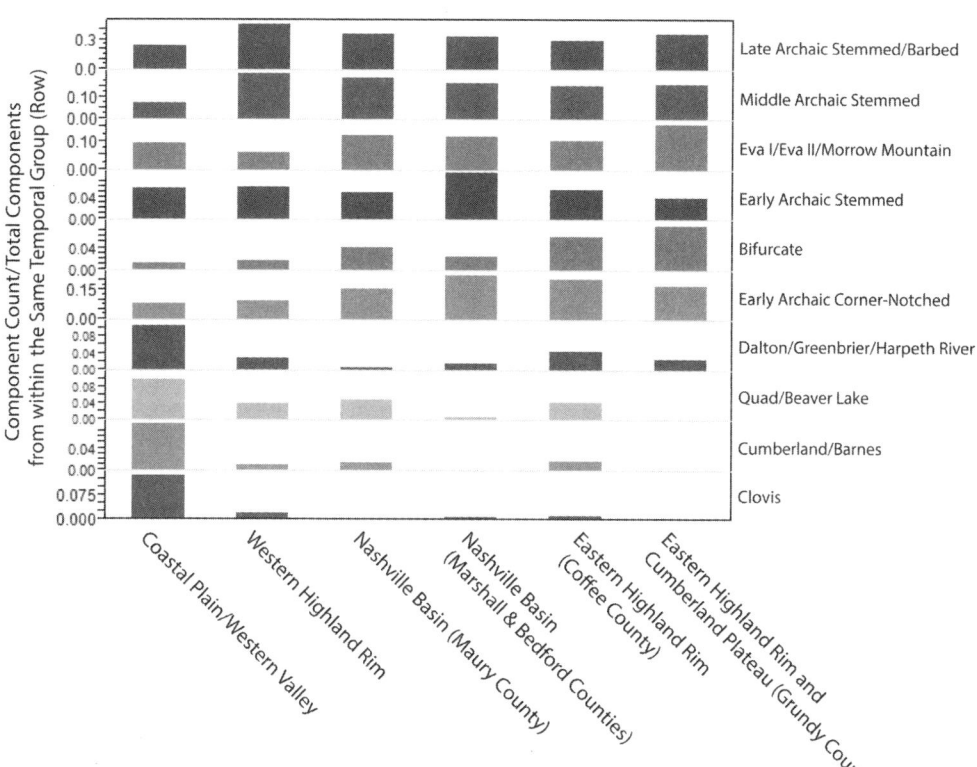

FIGURE 5.10. The distribution of temporal components by county divided by the total components from the same temporal group.

age, although fewer Eva I/Eva II/Morrow Mountain and Middle Archaic Stemmed components are reported in the Western Highland Rim.

I then standardized the study sample by dividing the number for each temporal component in each physiographic section by the total number of components from the same time period (Figure 5.10). For example, I divided the number of Clovis components reported in the Coastal Plain/ Western Valley by the total number of Clovis components represented in the drainage. Modifying the sample in this way helps to account for the taphonomic bias that affects the temporal distribution of sites by isolating the study sample at each time slice to analyze how each component varies across space. Again, three major trends are observable. First, the majority of the Paleoindian sites are found in the Coastal Plain/Western Valley physiographic section with increasing frequency in the Eastern Highland

TABLE 5.7. Results of chi-square analyses for the frequency of each temporal component compared to an even distribution of components in the Duck River drainage.

Component	n	X^2	df	p	V
Paleoindian	206	42.85	3	<0.001	0.32
Early Archaic Corner-Notched/Bifurcate	215	85.48	3	<0.001	0.45
Kirk Stemmed/Kirk Serrated/Stanly Stemmed	69	1.90	3	0.387	0.01
Eva I/Eva II/Morrow Mountain	118	17.30	3	<0.001	0.27
Middle Archaic Stemmed	180	15.36	3	<0.001	0.01
Late Archaic Stemmed/Barbed	387	10.79	3	0.005	0.12

Rim for the Quad/Beaver Lake and Dalton/Greenbrier/Harpeth River components. Similar to the previous analysis, both the Early Archaic Corner-Notched and Bifurcate components are more prevalent in the Eastern Highland Rim. Also similar to the previous analysis, beginning with the Early Archaic Stemmed, the subsequent components are more broadly distributed across the Duck River drainage, although there are fewer reported components in the Western Highland Rim for the Eva I/ Eva II/Morrow Mountain and Middle Archaic Stemmed components.

Finally, I used Pearson's goodness of fit test with Cramér's V as another means to illustrate the degree to which sites are distributed across the drainage (Table 5.7). As a null hypothesis, I again used an even distribution of sites across the drainage. To derive this, I divided the total number of components for each time slice by the percentage of the total area of the county. In other words, if a physiographic section encompasses 30 percent of the total area of the drainage, it should also contain 30 percent of the components for a given time slice if they were perfectly dispersed across the drainage. To meet the minimum sample size requirements for each cell, I combined the physiographic sections into three categories (Coastal Plain/Western Highland Rim, Nashville Basin, and Eastern Highland Rim/Cumberland Plateau). I also collapsed the Clovis, Cumberland, Quad/Beaver Lake, and Dalton/Greenbrier components into a general Paleoindian group and the Early Archaic Corner-Notched and Bifurcate components into another. This analysis illustrates that the trends described above are statistically significant. The Paleoindian and Early Archaic Corner-Notched/Bifurcate samples vary significantly from an even distribution, which is a result of the Paleoindian samples occurring more frequently in the Coastal Plain/Western Valley and the Early Archaic component in the Eastern Highland Rim.

While the remaining samples are significantly different from an even distribution of sites, their chi-square and Cramér's V values are much lower compared to the Paleoindian and Early Archaic Corner-Notched and Bifurcate distributions. In other words, while they differ significantly from an even distribution, they are much more evenly distributed than their Paleoindian and Early Archaic counterparts.

Discussion

Based on this analysis, there are several sources of variation in the reporting of sites in the Duck River drainage. This is partly due to differential reporting of sites across the drainage due to the frequency and extent of archaeological surveys. There is also a likely taphonomic bias that affects the recovery and reporting of progressively older sites (e.g. Surovell et al. 2009). To counteract the effects of these biasing factors, I divided each cell by both the total number of components for a given time slice, and again by the total number of Paleoindian and Archaic components from within the same physiographic section. Moreover, modern population density and the amount of stable upland surfaces relative to floodplain alluvium do not seem to impact the study sample.

From the analysis of bias-corrected data, there are three major trends in the distribution of sites. First, the Paleoindian period sites are predominantly found in the lower Tennessee River valley, although by the Late Paleoindian period, sites begin to appear in the Eastern Highland Rim. Second, starting with the Early Archaic Corner-Notched and Bifurcate components, sites are found more frequently in the Eastern Highland Rim. Third, with the appearance of the Early Archaic Stemmed components, sites are more evenly distributed across the physiographic provinces, although Eva I/Eva II/Morrow Mountain sites in the Western Highland Rim are slightly underrepresented.

Returning to the original hypotheses derived from the IFD set forth in the beginning of this chapter, the first prediction held in that the lower Tennessee River valley was the first habitat occupied. However, the second prediction is partially consistent with the archaeological data in that by the end of the Paleoindian period, the increased frequency of sites in the Eastern Highland Rim indicates that this was a secondary habitat that began to see increased use. But, loosely coinciding with the Pleistocene/Holocene transition, Early Archaic Corner-Notched and Bifurcate components began to appear more frequently in the Eastern Highland Rim, whereas the proportion diminished to the point where Bifurcate components are comparatively rare in the lower Tennessee River valley. The IFD model predicts that if the lower Tennessee River valley was still the highest-ranked habitat, it should continue to remain occupied, yet the distribution of archaeological sites appears to suggest that this was no longer the case during the Early Holocene.

This shift in focus from the lower Tennessee River valley to the Eastern Highland Rim over the course of the Younger Dryas is perhaps not that surprising given that the structure of the forests in eastern North America were fairly dynamic at a millennial scale and were responding to a warmer climate (e.g., Delcourt and Delcourt 1983). Paleoindian and Early Archaic groups were able to move into the Eastern Highland Rim and the Cumberland Plateau as the boreal forests gave way to deciduous forests at increasingly higher elevations. However, it is still not clear why there are fewer sites in the lower Tennessee River valley during the Early Holocene.

The third prediction of the IFD, which is that the frequency of sites in the Nashville Basin should diminish during the mid-Holocene, is also not consistent with the archaeological record. At the transition from the Early to Middle Archaic, the Early Archaic Stemmed components are more evenly distributed across the Duck River drainage, even though the retraction of oak-history forests and the expansion of cedar glades during the mid-Holocene would have made the Nashville Basin less desirable than the surrounding physiographic provinces. Since the Nashville Basin and the rest of the Duck River drainage were occupied for the remainder of the mid-Holocene, the most likely explanation is that people occupied the entire drainage as a result of population growth. This also renders the fourth hypothesis of the IFD moot: individuals should not be expected to expand into the Nashville Basin after the mid-Holocene if the Nashville Basin was never abandoned to begin with.

Conclusion

Both Binford (1968) and Flannery (1969) argued that population pressure was a critical component in explaining the context for the origins of agriculture in the Near East. Flannery (1969), in particular, argued that plant and animal domestication came about within the context of a broad spectrum revolution. However, in eastern North America, Smith (2007, 2011) argued that while people gravitated toward the major river valleys during the mid-Holocene, there is little to no evidence for population pressure in the time periods leading up to the domestication of plants in eastern North America.

For the lower Tennessee and Duck River valleys, the IFD provides a way to generate hypotheses that can be used to extrapolate demographic trends from the distribution of archaeological sites. In these drainages, the overall pattern in the distribution of archaeological sites is consistent with 1) expansion into lower elevation habitats during the Bølling-Allerød, Younger Dryas, and Early Holocene, 2) a diminished presence in the lower Tennessee River valley during the Early Holocene, and 3) a presence in every habitat at the end of the Early Holocene and in every subsequent time slice. In other words, the pattern is consistent with increasing population growth and limited flexibility to move to alternative habitats as a response to changing climatic conditions, particularly at the onset of the mid-Holocene epoch. Moreover, this conclusion would also point to one of the key implications of the IFD: once suitability is equalized among all habitats, suitability declines across all occupied habitats with population growth. Consequently, these results provide evidence for increasing population pressure in the time periods leading up to the appearance of domesticated plants in eastern North America.

CHAPTER 6

A Boom-Bust Model
for the Origins of Agriculture
in Eastern North America

Introduction

The transition from foraging to farming was one of the most momentous changes in prehistory, affecting everything from health to climate change in subsequent generations (Kennett and Winterhalder 2006). The most recent was in eastern North America, where the timing and suite of domesticated plants was firmly established by the work of Volney Jones (1936), Gayle Fritz (1990, 1997), Gary D. Crites (1993), Richard A. Yarnell (1978), and especially Bruce D. Smith (1987, 1992, 2001, 2011). However, the cultural context during which the appearance of domesticated plants emerged is still open to debate. Smith (2011) argues that eastern North America is one example that fails to support models that rely on resource depletion and population pressure as causal factors. This critique can be seen as part of a larger rejection of models originally put forth by Binford (1968) and Flannery (1969) that use resource imbalance to explain why the transition to agriculture occurred where and when it did.

When compared to later time periods—especially the Mississippian palisaded villages or Woodland period Hopewellian earthworks—the Late Archaic in the mid-continent seems rather simple (e.g., Anderson and Sassaman 2012:66; Caldwell 1958; Smith 2011:S471). However, when viewed across a broader time span, there is evidence for increasing population pressure and an imbalance of resources in the periods leading up to the first appearance of domesticated plants. To demonstrate this, I first summarize Smith's (2011) most recent formulation of his floodplain weed hypothesis. I then integrate the results of Chapters 4 and 5 to argue that at the end of the Early Archaic period, there is an abrupt change in the organization of biface technology and landscape use that likely had ramifications for the rest of the Archaic period. Rather than a gradual response to stable environmental conditions after the mid-Holocene (e.g., Caldwell 1958; Smith 2011), the appearance of domesticated plants occurred within the context of a millennial scale boom-bust cycle in available resources. I conclude with some suggestions for future research that could prove valuable in further exploration of this topic at a much finer temporal resolution at larger spatial scales and a defense for using the formal models of human behavioral ecology to construct hypotheses that can be tested with archaeological data.

Smith's Floodplain Weed Theory

Bruce D. Smith's (1987, 1992, 2011) model for the inception of domesticated plants and the emergence of a suite of co-occurring crops is considered to be the preeminent model for the origins of agriculture in eastern North America (Anderson and Sassaman 2012; Gremillion 1996, 2004). In the most recent formulation of his model, Smith (2011) focuses more specifically on the environmental and social factors influencing groups that made the transition from hunting and gathering to food production. He first describes the seven sites where the earliest domesticates have been recovered and then expands his discussion to the Late Archaic period in the mid-continent (5000–3400 cal BP). Smith argues that the earliest sites where domesticates have been identified occur along secondary and tertiary tributaries of the Mississippi River, which supports a floodplain (as opposed to upland cave/rockshelter) origin for the inception of domesticated plants. He then links this phenomenon to changes in fluvial geomorphology that occurred during the mid-Holocene, whereby aggrading rivers created the meander belt topography typical of the region today. This in turn created more plant- and animal-rich backswamps at a time when upland resources were deteriorating. Smith argues that it is within this context that people began to incorporate the weedy, r-selected species that are adept at colonizing the disturbed sediments along alluvial floodplains.

For the remainder of the Archaic period—and particularly the Late Archaic period—Smith (2011) contends that the archaeological record is remarkable for its stability and common patterns in technology, subsistence, and social organization. First, chipped stone assemblages, an absence of pottery, and open fire or heated stone cooking technology dominate all of the sites from this time period. Second, settlement size and organization shows little evidence for social differentiation and that based on the mortuary assemblages, it appears that these sites were composed of extended family units and ownership of resource catchments. Third, deer and a suite of smaller birds and mammals were the primary prey species, whereas hickory, oak, and walnut were the primary collected items because they can be gathered, processed, and stored with minimal effort. Fourth, Smith focused on the evidence of regional interaction and trade networks primarily as vectors for the movement of domesticated plants around the region.

Smith uses his characterizations of the Late Archaic to argue that the groups responsible for the domestication of plants in eastern North America consisted of small-scale societies and simple farming units with little to no status differentiation. Moreover, he finds that the case for population pressure or resource imbalance as a causal factor has yet to be made in eastern North America and cites Cheryl P. Claassen's (1996) observation that many river valleys were not intensively occupied in the Late Archaic. Smith (2011:S482) then asks two questions: 1) Why do some areas seemingly remain empty during the Late Archaic? and 2) Why do the primary subsistence and technological trends remain stable over the Archaic period?

Based on the results from the preceding chapters, it is possible to evaluate these two questions. First, while Smith (2011) cites Claassen's (1996) observation that many areas of eastern North America do not appear to be intensively occupied, an alternative hypothesis is that there are still areas of the Mid-South that have yet to be extensively surveyed or have site files that have not been updated since they were initially recorded (Anderson 1996:159). In other words, just because there are places where no archaeological sites have been recorded does not necessarily mean that people were not there. Instead, it is more beneficial to focus on drainages that have been extensively surveyed and reported. The Duck River is one of those places and has been the focus of several large-scale surveys. It displays evidence for continued occupation across all physiographic zones from the end of the Early Archaic period into the Middle and Late Archaic periods.

Second, Smith (2011) contends that the trends in subsistence and technology remained stable over the Archaic period. Based on the analysis discussed in Chapter 4, substantial changes in the organization of biface design, manufacture, and discard are likely related to variation in the returns from hunting over the Holocene. These two analyses run counter to Smith's (2011) assertions that areas were unoccupied and subsistence trends were stable. In the following sections, I provide an alternative boom-bust model for contextualizing the emergence of domesticated plants in eastern North America.

An Early Archaic–Middle Archaic Boom

Almost 30 years ago, Brown and Vierra (1983) published their observations regarding the Archaic period occupations at the Koster site in Illinois. Of particular interest, they noted that the mid-Holocene components, dating to between 9,000 and 5,800 cal BP, contained multiple dense occupations, a higher number of recorded features, evidence for intensive nut and seed processing, and the exploitation of aquatic species. These changes prompted Brown and Vierra to ask, What happened in the Middle Archaic?—perhaps one of the most famous questions in the archaeology of eastern North America. They hypothesized that changes in fluvial geomorphology promoted the formation of species-rich backwater swamps that, coupled with mid-Holocene warming, prompted groups to shift toward using river valleys in a more sedentary way. This punctuated "pull" model provided a stark contrast to Caldwell's (1958) primary forest efficiency model, whereby hunter-gatherers gradually adapted to the eastern woodlands over the course of the Archaic period. The trends that Brown and Vierra observed at Koster have also been noted in many other areas of eastern North America and are thought to foreshadow the appearance of intensive shellfish processing, monumental architecture, exchange networks, and the domestication of indigenous plant species (Anderson and Sassaman 2012:104–107).

One key problem with exploring the changes in subsistence in the Mid-South in the periods before the appearance of domesticated plants is that sites with preserved organic remains are relatively rare (Dunnell

1990; Styles and Klippel 1996). However, from the data that are available, it appears that during the Late Pleistocene through the mid-Holocene, prehistoric groups began to focus their dietary choices on a narrower set of highly ranked resources. This trend is reflected in the floral and faunal remains, the organization of biface technology, and the distribution of archaeological sites.

Stephen Carmody (2010) analyzed Middle Archaic paleobotanical remains from Dust Cave and three other rockshelters in northern Alabama, building on previous analyses by Kandice D. Hollenbach (2007) of the Late Paleoindian and Early Archaic period components from these same sites. Hollenbach argued that the manner in which plants were used changed little between the Late Paleoindian and Early Archaic. However, increased intensity of nut exploitation is suggested at the sites by the recovery of mortars and nutting stones and—most notably, at Dust Cave—a dramatic increase in the recovery of hickory nutshell within the Kirk Stemmed component at the close of the Early Archaic. Carmody found that, after incorporating the results of his analysis, this trend continued with Middle Archaic assemblages. He argued that when viewed through the lens of the diet breadth model (e.g., MacArthur and Pianka 1966), hickory nuts provide a source of calories, proteins, and fats and should be one of the most highly ranked plant resources in terms of caloric return when processed using a smash-and-boil method. Overall, the trend in paleobotanical remains from this sample of sites indicates an increasing emphasis on more highly ranked plant items from the Late Paleoindian through Middle Archaic.

Dust Cave is also well known for its preservation of faunal remains (Walker 1998, 2007). One of the key findings of Renee Walker's research was that the species recovered from the Late Paleoindian components were exceptionally diverse. The most striking pattern was the high percentage of avian species relative to the components from later time periods. Moreover, from the Late Paleoindian through Middle Archaic, there was an increase in the proportion of deer remains relative to smaller, relatively hard to catch species such as birds and squirrels. Additionally, the frequency of turtle and tortoise remains also increased over this time period. The inhabitants of Dust Cave appear to have focused their efforts on the highest-ranked species in terms of overall caloric return after adjusting for search and handling costs. This pattern in faunal use is not specific to Dust Cave. Styles and Klippel (1996) examined nine faunal assemblages from across the Mid-South and found a similar trend—most notably, that the relative frequency of deer dramatically increases in the Middle Archaic but then diminishes in later time periods.

In Chapter 4, I presented the results from an analysis of Paleoindian and Archaic bifaces from Benton and Humphreys counties in the lower Tennessee River valley. These counties have some of the highest densities of recorded Paleoindian artifacts in eastern North America (Anderson et al. 2010; Broster et al. 2013), contain several key Archaic period type sites (e.g., Lewis and Kneberg 1959), and are located near the geographic center of the sites with the earliest dated domesticated plants in eastern North America (e.g., Smith and Yarnell 2009).

I found that Late Paleoindian and Early Archaic points were relatively smaller and displayed, on average, proportions consistent with extensive resharpening. However, a major break in this pattern occurs with the Eva I type, which are discarded with relatively large amounts of unexpended utility. Using the model developed by Kuhn and Miller (2015), bolstered with the analysis of faunal remains and technological organization from Puntutjarpa Cave (Gould 1977, 1980) and Gatecliff Shelter (Thomas 1983) in Chapter 3, I interpret this pattern as indicative of exceptionally higher returns from hunting. This finding is consistent with the faunal record at Dust Cave (Walker 1998, 2007) and the sample of sites across the Mid-South analyzed by Styles and Klippel (1996), where the frequency of deer to smaller mammal remains increases substantially from the Early to Middle Archaic.

This break in the flora, fauna, and biface technological organization also correlates with a major shift in the distribution of archaeological sites. During the Paleoindian period, sites in the Duck River drainage are more frequently found near the confluence with the lower Tennessee River, but by the Late Paleoindian period, sites begin to appear in the Eastern Highland Rim. Then, during the Early Archaic, the frequency of sites shifts disproportionately to the Eastern Highland Rim. At the end of the Early Archaic period, sites are found across the entirety of the drainage. I interpret this pattern as indicating that the Paleoindian and Early Archaic populations were low enough that they had both the flexibility to move and access to the most productive habitats as a means to adjust to changing environmental conditions. However, by the end of the Early

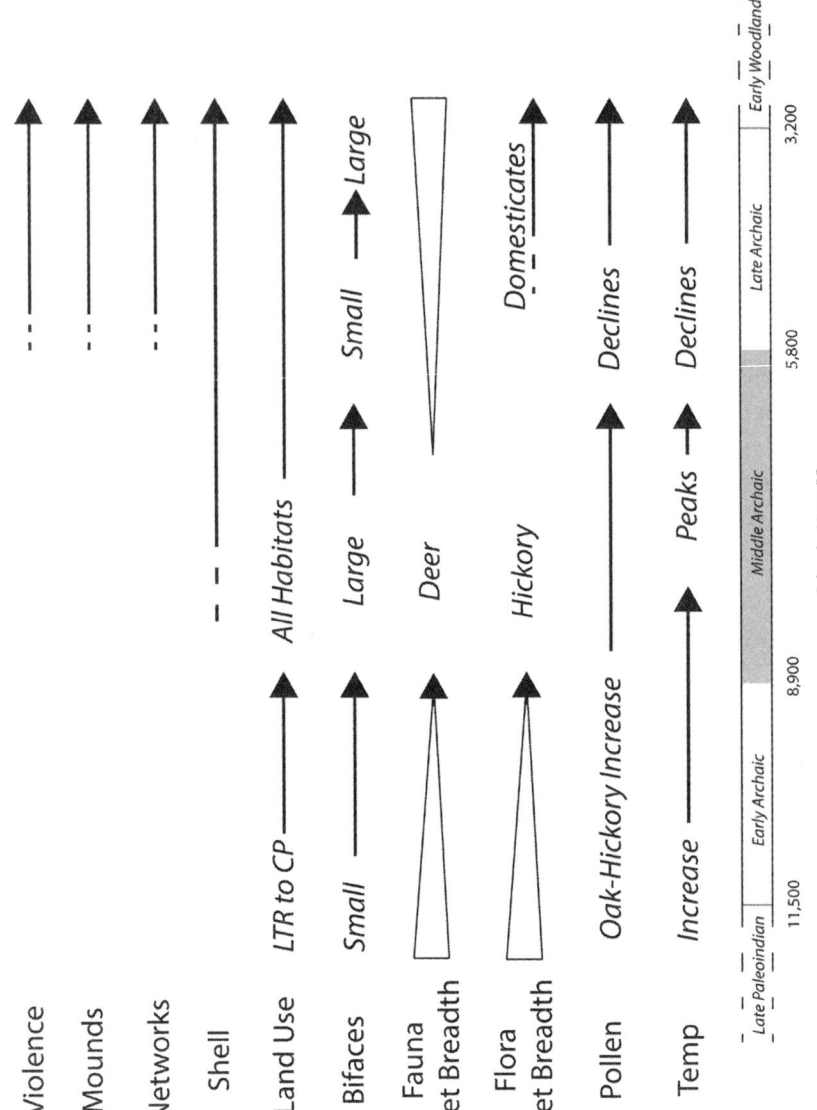

FIGURE 6.1. Major environmental, subsistence, technological, land use, and cultural trends during the Paleoindian and Archaic periods in the Mid-South (LTR— Lower Tennessee River; CP—Cumberland Plateau).

Archaic, loosely coinciding with the increasing focus on highly ranked resources at the beginning of the mid-Holocene, populations were found throughout the drainage.

Like Brown and Vierra (1983), I would argue that climate change played a prominent role in the observed changes in subsistence. While the mid-Holocene is generally considered a period of warmer, drier conditions relative to modern climate, ongoing research from a variety of sources has identified substantial regional variation. An increase in temperature over the Early Holocene, which peaked in roughly 7,000 cal BP, would have greatly altered the resources available for prehistoric groups. The pollen cores from both Jackson Pond in Kentucky (Wilkins et al. 1991) and Anderson Pond (Delcourt 1979) in central Tennessee show large increases in oak pollen at this time. Moreover, the rise in temperature over the course of the Holocene likely would have increased the masting productivity and frequency of bumper crops in oak- and hickory-dominated forests (Bissett 2010; Gardner 1997).

The changing environmental conditions during the mid-Holocene would have had two major ramifications for hunter-gatherer groups. First, nearly 90 animal species are known to feed on acorns; deer represent one of the most significant acorn consumers in modern oak forests. During years of high acorn productivity, as much as 83 percent of the diet of white-tailed deer is comprised of acorns, and these boom periods are associated with higher average female body weight, increased frequency of ovulation, higher birthrates, and higher birth weights of fawns (Gardner 1997; Bissett 2010). Consequently, high masting productivity, which probably characterized the peak years of the mid-Holocene, would have resulted in greater health and fertility in deer populations.

Second, rather than competing with deer for acorns, the increased availability of hickory nuts would have provided an alternative, more highly ranked food source for humans to intensively exploit. Consequently, the expansion of oak-hickory forests during the Early and mid-Holocene likely led to a millennial-scale boom in resources capped by an additional peak in temperature 7,000 years ago. In other words, the intensive focus on deer and hickory in the Middle Archaic was the result of a wave of abundance that began gathering momentum at the beginning of the Holocene (Figure 6.1).

A Middle Archaic–Late Archaic Bust

Moore and Dekle (2010) argued that at the end of the Early Archaic and the beginning of the Middle Archaic periods, a radical shift in household economies occurred across the Mid-South that included a focus on more stable, predictable, and immobile resources, such as hickory and shellfish. Consequently, this shift in household economies allowed for greater food storage capabilities, which Gremillion (2004) contends was the major innovation that allowed hunter-gatherers in the Mid-South to delay consumption and bank calories to overcome the late winter/early spring shortfalls.

However, the optimal conditions did not persist through the duration of the Middle Archaic period. Based on the pollen sequence at Anderson Pond (Delcourt 1979), the relative abundance of oak and hickory diminished, while at Cheek Bend Cave (Klippel and Parmalee 1982), there is a shift in small mammal species indicative of a retreat in the distribution of upland glades that had expanded during the peak of the mid-Holocene. Based on the projection by Viau and colleagues (2006) and the vegetation reconstructions by Williams and colleagues (2004) and Delcourt and Delcourt (1983, 1985), mean annual temperatures decreased and the oak-hickory dominated forests continued to move north as a result of increasing moisture and the decreasing mean July temperature.

The shift in climate and environmental conditions would have impacted the returns from both hunting and gathering over the long term. For hunting, this decline in returns is evident in the use of bifaces with less overall mass being invested in the initial production of Eva II/Morrow Mountain and Middle Archaic Stemmed points. I argue that this indicates declining returns from hunting, especially when compared to the Early Archaic Stemmed and Eva I groups. This trend correlates with the decline in deer relative to smaller mammals in archaeological assemblages noted by Styles and Klippel (1996) and Bissett (2010) that begins in the Middle Archaic. One consequence of declining hunting returns was that the distance and duration of hunting forays would have also likely increased. However, the distribution of sites across the entire drainage indicates that there were relatively few, if any, unoccupied areas in the drainage at this time. In other words, with the decline of deer, acorns, and hickory nuts and the absence of mobility options restricting movement, individuals began to focus more intensively on extracting limited resources from more limited areas and also creating more rigid territories.

Rather than expanding their diet breadth to encompass a wider range of resources as hickory and deer diminished, it appears that some groups began to focus more intensively on weedy annuals, including some that would eventually become domesticated. Doing so would have allowed some groups to expand the productivity of their resource catchments by finding alternative items to fill their storage pits as the availability of hickory and acorns waned, an example of hunter-gatherer groups innovating to expand productivity in a limited area.

In a context such as the one just described, there should be increasing evidence for groups defining and defending territories (Dyson-Hudson and Smith 1978). Based on ethnographic research with four hunter-gatherer groups in the Kalahari Desert, Cashdan (1983) argued that there are two ways to manage unexpected visitors to one's territory. First, create rules for granting access (e.g., social boundary defense), which is usually facilitated by gift exchange. This is visible archaeologically in the Middle Archaic when exchange networks begin to emerge—most notably, with the exchange of bone pins in the Ohio River, Benton Interaction Sphere in the lower Mid-South and other items in eastern North America (Jefferies 1996, 2004). In the case of the Benton bifaces, the exchange of well-crafted, large bifaces contrasts with the broader pattern of smaller, more expediently discarded bifaces during the Middle Archaic, which probably helped to underwrite their value as an exchange item in the Mid-South.

Cashdan (1983) identified other contexts where an unexpected intrusion could provide grounds for a violent response (e.g., perimeter defense). In the Mid-South, human skeletal remains found in the Tennessee, Cumberland, Duck, and Green River valleys, partially as a result of the excellent preservation of remains found in shell mounds and middens, provide evidence of these sorts of more antagonistic interactions. When compared to skeletal samples from across eastern North America, the individuals from Middle and Late Archaic sites, including several in the lower Tennessee River valley that were used in the study sample in Chapter 4, show evidence for trauma and trophy taking (scalping and/or the removal of limbs and crania) indicative of intergroup

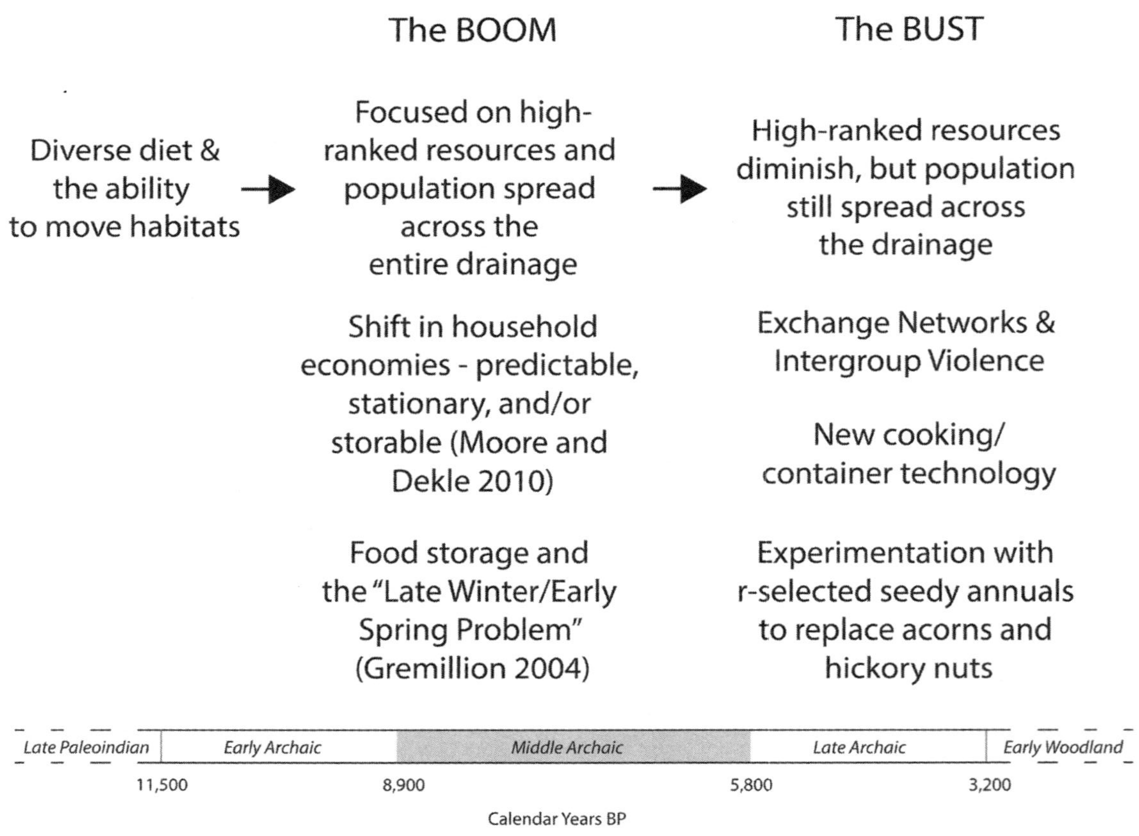

The BOOM

The BUST

Diverse diet &
the ability →
to move habitats

Focused on high-
ranked resources and
population spread
across the
entire drainage

→

High-ranked resources
diminish, but population
still spread across
the drainage

Shift in household
economies - predictable,
stationary, and/or
storable (Moore and
Dekle 2010)

Exchange Networks &
Intergroup Violence

New cooking/
container technology

Food storage and
the "Late Winter/Early
Spring Problem"
(Gremillion 2004)

Experimentation with
r-selected seedy annuals
to replace acorns and
hickory nuts

Late Paleoindian	Early Archaic	Middle Archaic	Late Archaic	Early Woodland
11,500	8,900	5,800	3,200

Calendar Years BP

FIGURE 6.2. A boom-bust model for the origins of agriculture in the Mid-South.

violence and competition (Mensforth 2007; Milner 1999; Smith 1995, 1996). Consequently, after the boom in resources and population in the mid-Holocene, there is evidence for the emergence of more rigid, re-gionalized territories in the Mid-South, rules for entering adjacent terri-tories with permission, and examples for what happens if you get caught trespassing.

Plant Domestication and Behavioral Ecology

In the opening chapter, I argue that the formal models of human behav-ioral ecology (e.g., Kennett and Winterhalder 2006; Smith et al. 2001) could be used to diagnose and interpret major trends in subsistence over time. This study is, in effect, a deep history macroeconomic approach (e.g., Stiner and Feeley-Harnik 2011) that examines the environmental and demographic context prior to the appearance of domesticated plants in eastern North America.

Based on this study, as well as supporting lines of evidence, there are several major trends in subsistence that have ramifications for contextu-alizing the emergence of the suite of domesticated plants known as the Eastern Agricultural Complex (Smith 2001; Smith and Yarnell 2009; Fig-ure 6.2). Based on flora and faunal remains, as well as the organization of biface technology and the distribution of archaeological sites, Late Paleo-indian diet breadth was relatively diverse, yet population density was low enough that groups had the flexibility to move around the landscape to respond to changes in environmental conditions at a very gross scale. During the Early Holocene and mid-Holocene, however, these trends reversed, with an increasing reliance on high-ranked plant and animal species. The shift in economic focus is likely related to environmental changes, including a broad warming trend in the Early Holocene that is associated with an increase in oak and hickory pollen and an increasing abundance and diversity of species. The outcome was an increase in the availability of hickory nuts (the highest-ranked gathered item for people) and acorns (the preferred food item for deer). This trend culminated in a dramatic shift in household economies at the end of the Early Ar-chaic and the beginning of the Middle Archaic (e.g., Moore and Dekle 2010) to a reliance on more stable, predictable resources, which included shellfish in addition to nut mast. In other words, it was in the context of abundance, not scarcity that the Shell Mound Archaic began to gain momentum in the Mid-South.

The shift in household economies had several unintended consequences for subsequent generations. First, it created the context for decreasing residential mobility as abundant, predictable resources allowed hunter-gatherers to procure sufficient calories. Second, the reduction in mobility in the context of abundance allowed groups to exploit resources suitable for food storage and select places to store these resources. Third, the seasonal gap in resources in the late winter/early spring provides a microeconomic catalyst to find ways in which to bank calories and delay consumption as a buffer against anticipated shortfalls (Gremillion 2004; Woodburn 1982). Finally, the increase in the abundance of highly ranked resources, the decrease in mobility, and the ability to bank calories to offset shortfalls likely led to a population boom in the Mid-South, reflected locally in the widespread distribution of sites across the Duck River drainage and the discard of projectile points with a high degree of expended utility, something we would only expect if the returns from hunting were exceptionally high.

However, these conditions did not persist through the duration of the mid-Holocene. The warm, dry conditions that drove the abundance of oak and hickory masting receded, which is reflected locally by a reduction in oak and hickory pollen in Anderson Pond and the species composition found in Cheek Bend Cave (Delcourt 1979; Klippel and Parmalee 1982). As the conditions that prompted the boom unraveled, several factors prevented local hunter-gatherer groups from simply diversifying their diet breadth and increasing the frequency of residential moves to acquire sufficient resources. First, by this point, groups display evidence for an increasing reliance on stored resources (Moore and Dekle 2010). Second, there were fewer unoccupied habitats, reflected in the continued widespread distribution of sites across the entire Duck River drainage.

Consequently, even if people were willing to give up the source and location of their stored calories, the presence of other groups in these locations would limit their options for moving and exploiting alternative habitats. Such potential interference would increase the costs of giving up one habitat and moving to another. Instead of moving, groups invested in technology that would allow them to extract more calories from more limited resource catchments, which include the use of ground stone and other innovations to extract more calories from gathered goods, mortar holes on the Cumberland Plateau (Franklin 2002), soapstone, pottery, and associated cooking technologies (Sassaman 1993). They also began experimenting with seedy annuals that could be used to fill storage pits in lieu of acorn and hickory nuts. Restrictions on mobility and a greater investment in extracting food value from limited resources also led to an emphasis on defining and defending territories, reflected in the emergence of large multigenerational cemeteries (Smith 1996), continued production of shell mounds and middens (Claassen 1996; Marquardt and Watson 2005), regional exchange networks (Jefferies 1996, 2004), and increasing evidence for intergroup violence and warfare (Mensforth 2007; Milner 1999; Smith 1996) during the Middle and Late Archaic in the Mid-South. Rather than plant domestication in eastern North America emerging in the context of abundance (e.g., Hayden 1992, 1995; Smith 2011), it came as the result of a sequential boom-bust cycle rooted in changing Holocene environmental conditions and increasing population pressure (e.g., Binford 1968; Flannery 1969).

Future Directions

Both Smith (1987, 1992, 2001, 2011) and Gremillion (1996, 2004) observe that plant domestication in eastern North America is a phenomenon that occurred not as a single event at a single place but as a culmination of processes that occurred on a regional scale. For this study, I integrated these various sources of information in a much more limited area to argue that in the Duck River drainage, the appearance of domesticated plants occurred within the context of a macroeconomic boom-bust cycle in available resources. By increasing the size and scale to a truly regional level, it will be possible to more fully explore whether the patterns observed in the Duck River are evident elsewhere as part of an explanation for the origins of agriculture in eastern North America. Consequently, there is a need for datasets that can meet the requirements for exploring the context of plant domestication in eastern North America.

One such dataset that would be immediately valuable is to systematically collect and consolidate archaeological radiocarbon dates into a single database. Successful examples include the Canadian Archaeological Radiocarbon Database (Morlan 2005) and the database of archaeological radiocarbon dates from Australia (Williams 2013). Most information on available radiocarbon dates from eastern North America can be found only in summary articles or regional gray literature, often with minimal information on the context from which they came and the artifacts with which they were associated (Miller and Gingerich 2013b).

Second, the chronological framework for eastern North America relies heavily on temporally diagnostic bifaces from a relatively limited number of type sites. One model for streamlining information on this artifact class is the Paleoindian Database of the Americas (PIDBA), an ongoing project that aims to integrate artifact images, qualitative and metric attributes, and locational information on Paleoindian-aged artifacts from North and South America (Anderson et al. 2010). More specifically, rather than just images of the ideal types and average metric attributes such as those found in Justice (1995), a database that provides the full range of variation from key sites could provide the basis for more robust comparisons when coupled with approaches that make use of geometric morphometrics (e.g., Buchanan 2006; Thulman 2012).

Third, there are several studies I have repeatedly cited that have demonstrated the value of compiling and integrating data at large scales to examine broad trends in subsistence. Moreover, there has been a call for integrating these datasets, specifically in eastern North America (Hollenbach and Walker 2010; VanDerwarker and Peres 2010). Given that sites with preserved organic remains are more difficult to find with increasing antiquity, a premium should be placed on reporting and compiling this type of information. Two effective examples are FAUNMAP (Graham and Lundelius 2010) and the North American Pollen Database (COHMAP Members 1988).

In Chapter 5, I argued that site file records can be useful for interpreting land-use trends to make inferences concerning demography and referenced Anderson's (1996) study of site distributions across the lower Southeast. More recently, a project is underway to fully integrate the site file records in eastern North America in a single database (Anderson et al. 2012). One future consideration for such an undertaking would be to include the full information on how temporal designations are made to allow analysts to further combine or subdivide sites temporally beyond major culture-historical units. Considering the effect of the transition to agriculture on modern societies, and that eastern North

America is one of the few places where we can study how this came about, integrating large datasets to pursue this big question beyond the lower Tennessee and Duck River valleys would be well worth the effort.

Finally, I would like to conclude by echoing the sentiments of Gremillion (2004), Gremillion and Piperno (2009); Gremillion and colleagues (2014a, 2014b, 2014c) Kennett and Winterhalder (2006), Stiner (2001), and others who advocate the use of behavioral ecology to understand the archaeological record (Bettinger 2009; O'Connell 1995; Bird and O'Connell 2006; Codding and Bird 2015; Kelly 2013; Surovell 2009). While there have been many critiques of the assumptions surrounding economic rationality and human behavior, I would argue that the formal models of behavioral ecology provide valuable points of departure for examining the archaeological record. In this particular study, these models allowed me to construct testable predictions; they guided the creation and analysis of large datasets to test those predictions and helped generate alternative hypotheses to explain patterns in the archaeological record. Moreover, these models also allowed me to incorporate two datasets—stone tools and the distribution of archaeological sites—to examine the origins of agriculture, a question that has thus far been debated between those with backgrounds in paleobotany and zooarchaeology. I am by no means claiming to reinvent the wheel, but I instead join the chorus of those advocating for the utility of human behavioral ecology to explore the origins of agriculture.

Bibliography

Abbott, Charles C.

1877　On the Discovery of Supposed Paleolithic Implements from the Glacial Drift, in the Valley of the Delaware River, near Trenton, New Jersey. Tenth Annual Report of the Trustees of the Peabody Museum of American Archaeology and Ethnology Vol. 2, No. 1. Harvard University, Cambridge, MA.

1889　Evidences of the Antiquity of Man in Eastern North America. *Proceedings of the American Association for the Advancement of Science* 37:293–315.

Alley, Richard B.

2000　Ice-Core Evidence of Abrupt Climate Changes. *Proceedings of the National Academy of Sciences* 97(4):1331–1334.

Amick, Daniel S.

1987　*Lithic Raw Material Variability in the Central Duck River Basin: Reflections of Middle and Late Archaic Organizational Strategies.* Tennessee Valley Authority, Cultural Resources Program, Knoxville.

1996　Regional Patterns of Folsom Mobility and Land Use in the American Southwest. *World Archaeology* 27(3):411–426.

Amick, Daniel S., and Phillip Carr

1996　Changing Strategies of Lithic Technological Organization. In *Archaeology of the Mid-Holocene Southeast*, edited by Kenneth E. Sassaman and David G. Anderson, pp. 41–56. University Press of Florida, Gainesville.

Anderson, David G.

1996　Approaches to Modeling Regional Settlement in the Archaic Period Southeast. In *Archaeology of the Mid-Holocene Southeast*, edited by Kenneth E. Sassaman and David G. Anderson, pp. 157–176. University Press of Florida, Gainesville.

2001　Climate and Culture Change in Prehistoric and Early Historic Eastern North America. *Archaeology of Eastern North America* 29:143–186.

2005   Pleistocene Human Occupation of the Southeastern United States: Re-
        search Directions for the Early 21st Century. In *Paleoamerican Origins:
        Beyond Clovis*, edited by Robson Bonnichsen, Bradley T. Lepper, Den-
        nis Stanford, and Michael R. Waters, pp. 29–42. Texas A&M University
        Press, College Station.

Anderson, David G., and J. Christopher Gillam
2000   Paleoindian Colonization of the Americas: Implications from an
        Examination of Physiography, Demography, and Artifact Distribution.
        American Antiquity 65(1):43–66.

Anderson, David G., and Glen T. Hanson
1988   Early Archaic Settlement in the Southeastern United States: A Case
        Study from the Savannah River Valley. *American Antiquity* 53(2):
        262–286.

Anderson, David G., and Kenneth E. Sassaman
2012   *Recent Developments in Southeastern Archaeology: From Colonization to
        Complexity*. Society for American Archaeology Press, Washington, DC.

Anderson, David G., Eric Kansa, Sarah Kansa, Stephen Yerka, and Joshua Wells
2012   *Developing the Cyberinfrastructure for a National Archaeological Site
        Database*. Proposal submitted to the National Science Foundation,
        Arlington, VA.

Anderson, David G., D. Shane Miller, Stephen J. Yerka, J. Christopher Gillam,
Erik N. Johanson, Derek T. Anderson, Albert C. Goodyear, and Ashley M.
Smallwood
2010   PIDBA (Paleoindian Database of the Americas) 2010: Current Status
        and Findings. *Archaeology of Eastern North America* 38:1–18.

Anderson, David G., Michael R. Russo, and Kenneth E. Sassaman
2007   Mid-Holocene Cultural Dynamics in Southeastern North America. In
        *Climate Change and Cultural Dynamics: A Global Perspective on Mid-
        Holocene Transitions*, edited by David G. Anderson, Kirk A. Maasch,
        and Daniel H. Sandweiss, pp. 457–490. Elsevier Press, London.

Andrefsky, William, Jr.
1994   Raw-Material Variability and the Organization of Technology. *Ameri-
        can Antiquity* 59 (1):21–34.
2006   Experimental and Archaeological Verification of an Index of Retouch
        for Hafted Bifaces. *American Antiquity* 71(4):743–757.
2009   The Analysis of Stone Tool Procurement, Production, and Mainte-
        nance. *Journal of Archaeological Research* 17:65–103.

Angst, Michael G., Bradley A. Creswell, Matthew D. Gage, Gail G. Guymon,
and Stephen J. Yerka
2011   *Archaeological Survey of TVA Lands along the Lower Duck River and*

within the Duck River and Big Sandy Units of the Tennessee National Wildlife Refuge, Benton, Henry, and Humphreys Counties, Tennessee. Prepared for Tennessee Valley Authority Cultural Compliance Section. Archaeological Research Laboratory, Department of Anthropology, University of Tennessee, Knoxville.

Archer, Michael

1977   Faunal Remains from the Excavation at Puntutjarpa Rockshelter. In *Puntutjarpa Rockshelter and the Australian Desert Culture*, edited by Richard Gould, pp. 158–165. Anthropological Papers Vol. 54, Pt. 1. American Museum of Natural History, New York.

Asch, David L., and Nancy B. Asch

1985   Prehistoric Plant Cultivation in West-Central Illinois. In *Prehistoric Food Production in North America*, edited by Richard I. Ford. Museum of Anthropology, University of Michigan, Ann Arbor.

Asch, Nancy B., Richard I. Ford, and David L. Asch

1972   *Paleoethnobotany of the Koster Site: The Archaic Horizons.* Reports of Investigations No. 24, Illinois State Museum, Springfield.

Ballenger, Jesse A. M., and Jonathan B. Mabry

2011   Temporal Frequency Distributions of Alluvium in the American Southwest: Taphonomic, Paleohydraulic, and Demographic Implications. *Journal of Archaeological Science* 38(6):1314–1325.

Bamforth, Douglas B.

1986   Technological Efficiency and Tool Curation. *American Antiquity* 51(1): 38–50.

Barker, Graeme

2006   *The Agricultural Revolution in Prehistory.* Oxford University Press, Oxford.

Beck, Charlotte, Amanda K. Taylor, George T. Jones, Cynthia M. Fadem, Caitlyn R. Cook, and Sara A. Millward

2002   Rocks Are Heavy: Transport Costs and Paleoarchaic Quarry Behavior in the Great Basin. *Journal of Anthropological Archaeology* 21:481–507.

Bennett, Deborah, J.

1998   *Randomness.* Harvard University Press, Cambridge, MA.

Bettinger, Robert L.

2009   *Hunter-Gatherer Foraging: Five Simple Models.* Eliot Werner, Clinton Corners, NY.

Bettinger, Robert L., and Jelmer Eerkens

1999   Point Typologies, Cultural Transmission, and the Spread of Bow-and-Arrow Technology in the Prehistoric Great Basin. *American Antiquity* 64(2):231–242.

Binford, Lewis R.

1968   *Post-Pleistocene Adaptations*. In *New Perspectives in Archeology*, edited
       by Sally R. Binford and Lewis R. Binford, pp. 313–341. Aldine, Chicago.

1977   Introduction. In *For Theory Building in Archaeology: Essays on Faunal
       Remains, Aquatic Resources, Spatial Analysis, and Systemic Modeling*,
       pp. 1–10. Academic Press, New York.

1978   *Nunamiut Ethnoarchaeology*. Academic Press, New York.

1979   Organization and Formation Processes: Looking at Curated Tech-
       nologies. *Journal of Anthropological Research* 35(3):255–273.

1980   Willow Smoke and Dogs' Tails: Hunter-Gatherer Settlement Systems
       and Archaeological Site Formation. *American Antiquity* 45(1):4–20.

1981   Behavioural Archaeology and the "Pompeii Premise." *Journal of Anthro-
       pological Research* 37:195–208.

2001   *Constructing Frames of Reference: An Analytical Method for Archaeolog-
       ical Theory Building Using Ethnographic and Environmental Data Sets*.
       University of California Press, Berkeley.

Bird, Douglas W., and James F. O'Connell

2006   Behavioral Ecology and Archaeology. *Journal of Archaeological Research*
       14(2):143–188.

Bissett, Thaddeus

2010   "Linking Resource Abundance, Population, and the Rise of Regional
       Exchange Networks in the Middle Archaic Midsouth and Lower Mid-
       west." Paper presented at the 67th Annual Meeting of the Southeastern
       Archaeological Conference, Lexington, KY.

2014   "The Western Tennessee Shell Mound Archaic: Prehistoric Occupation
       in the Lower Tennessee River Valley between 9000 and 2500 cal yr BP."
       Ph.D. dissertation, Department of Anthropology, University of Tennes-
       see, Knoxville.

Bissett, Thaddeus G., and D. Shane Miller

2017   "Refining the Ages of Paleoindian through Terminal Late Archaic Types
       in the Lower Midsouth Using Bayesian Statistical Modeling." Poster
       presented at the Current Research in Tennessee Archaeology Annual
       Meeting in Nashville.

Bleed, Peter

1986   The Optimal Design of Hunting Weapons: Maintainability or Reliabil-
       ity. *American Antiquity* 51(4):737–747.

Bliege Bird, Rebecca, and Eric A. Smith

2005   Signaling Theory, Strategic Interaction, and Symbolic Capital. *Current
       Anthropology* 46:221–248.

Blondel, Jacques

2006   The "Design" of Mediterranean Landscapes: A Millennial Story of

Humans and Ecological Systems during the Historic Period. *Human Ecology* 34(5):713–729.

Boldurian, Anthony T., and John L. Cotter

1999 *Clovis Revisited: New Perspectives on Paleoindian Adaptations from Blackwater Draw, New Mexico.* University Museum Monograph No. 103. University of Pennsylvania Press, Philadelphia.

Borgerhoff Mulder, Monique, and Ryan Schacht

2012 Human Behavioural Ecology. In *Encyclopedia of Life Sciences*, pp. 1–10, Wiley: Chichester.

Box, George E. P., and Norman R. Draper

1987 *Empirical Model-Building and Response Surfaces.* John Wiley and Sons, New York.

Brackenridge, G. Robert

1984 Alluvial Stratigraphy and Radiocarbon Dating along the Duck River, Tennessee: Implications Regarding Floodplain Origin. *Geological Society of America Bulletin* 95(1):9–25.

Braidwood, Robert J.

1960 The Agricultural Revolution. *Scientific American* 203:130–148.

Brantingham, P. Jeffrey

2006 Measuring Forager Mobility. *Current Anthropology* 47(3):435–459.

Breitburg, Emmanuel, and John B. Broster

1994 Paleoindian Site, Lithic, and Mastodon Distributions in Tennessee. *Current Research in the Pleistocene* 11:9–11.

1995 Clovis and Cumberland Projectile Points of Tennessee: Quantitative and Qualitative Attributes and Morphometric Affinities. *Current Research in the Pleistocene* 12:4–6.

Bronk Ramsey, Christopher

2008 Radiocarbon Dating: Revolutions in Understanding. *Archaeometry* 50(2):249–275.

2009 Bayesian Analysis of Radiocarbon Dates. *Radiocarbon* 51(1):337–360.

Bronk Ramsey, Christopher, and Sharen Lee

2013 Recent and Planned Developments of the Program OxCal. *Radiocarbon* 55(2–3):720–730.

Broster, John B.

1989 A Preliminary Survey of the Paleo-Indian Sites in Tennessee. *Current Research in the Pleistocene* 6:29–31.

Broster, John B., and Mark R. Norton

1990 Lithic Analysis and Paleo-Indian Utilization of the Twelkemeier Site (40HS173). *Tennessee Anthropologist* 15:115–131.

1991 Clovis and Cumberland Sites in the Kentucky Lake Region. *Current Research in the Pleistocene* 8:10–12.

1993 The Carson-Conn-Short Site (40BN190): An Extensive Clovis Habitation in Benton County, Tennessee. *Current Research in the Pleistocene* 10:3–5.

1996 Recent Paleoindian Research in Tennessee. In *The Paleoindian and Early Archaic Southeast*, edited by David G. Anderson and Kenneth E. Sassaman, pp. 288–297. University of Alabama Press, Tuscaloosa.

Broster, John B., Mark R. Norton, D. Shane Miller, Jesse W. Tune, and Jonathan Baker

2013 Tennessee's Paleoindian Record: The Cumberland and Lower Tennessee River Watersheds. In *In the Eastern Fluted Point Tradition*, edited by Joseph A. M. Gingerich, pp. 299–314. University of Utah Press, Salt Lake City.

Brown, James A., and Robert K. Vierra

1983 What Happened in the Middle Archaic? Introduction to an Ecological Approach to Koster Site Archaeology. In *Archaic Hunters and Gatherers in the American Midwest*, edited by James L. Phillips and James A. Brown, pp. 165–196. Academic Press, New York.

Broyles, Bettye J.

1966 Preliminary Report: The St. Albans Site (46KA27), Kanawha County, West Virginia. *West Virginia Archaeologist* 19:1–43.

Buchanan, Briggs

2003 The Effects of Sample Bias on Paleoindian Fluted Point Recovery in the United States. *North American Archaeologist* 24(4):311–338.

2006 An Analysis of Folsom Projectile Point Resharpening Using Quantitative Comparisons of Form and Allometry. *Journal of Archaeological Science* 33:185–199.

Buchanan, Briggs, Mark Collard, and Kevan Edinborough

2008 Paleoindian Demography and the Extraterrestrial Impact Hypothesis. *Proceedings of the National Academy of Sciences* 105(33):11651–11654.

Buchanan, Briggs, Mark Collard, Marcus J. Hamilton, and Michael J. O'Brien

2011 Points and Prey: A Quantitative Test of the Hypothesis that Prey Size Influences Early Paleoindian Projectile Point Form. *Journal of Archaeological Science* 38:852–864.

Bullen, Ripley P.

1975 *A Guide to the Identification of Florida Projectile Points.* Kendall Books, Gainesville, FL.

Burger, Oskar, Marcus J. Hamilton, and Robert Walker

2005 The Prey as Patch Model: Optimal Handling of Resources with Diminishing Returns. *Journal of Archaeological Science* 32(8):1147–1158.

Byers, David A., and Jack M. Broughton

2004 Holocene Environmental Change, Artiodactyl Abundances, and

Human Hunting Strategies in the Great Basin. *American Antiquity*
69(2):235–255.

Byers, David A., and Craig S. Smith

2007   Ecosystem Controls and the Archaeofaunal Record: An Example from
the Wyoming Basin, USA. *The Holocene* 17(8):1171–1183.

Cabak, Melanie A., Kenneth E. Sassaman, and J. Christopher Gillam

1996   *Distributional Archaeology in the Aiken Plateau: Intensive Survey of
E Area, Savannah River Site, Aiken County, South Carolina.* Savannah
River Archaeological Research Papers No. 8. Occasional Papers of the
Savannah River Archaeological Research Program, South Carolina
Institute of Archaeology and Anthropology, University of South Caro-
lina, Columbia.

Caldwell, Joseph R.

1958   *Trend and Tradition in the Prehistory of the Eastern United States.*
American Anthropological Association Memoir No. 88. American
Anthropological Association, Menasha, WI.

Cambron, James W., and David C. Hulse

1969   *Handbook of Alabama Archaeology: Part I, Point Types.* Archaeological
Research Association of Alabama, Birmingham.

1975   *Handbook of Alabama Archaeology, Part I, Point Types.* Revised ed.
Archaeological Research Association of Alabama, Tuscaloosa.

Carmody, Stephen

2009   "Hunter/Gatherer Foraging Adaptations during the Middle Archaic
Period at Dust Cave, Alabama." Master's thesis, Department of Anthropology, University of Tennessee, Knoxville.

2010   "The Relationship Between Middle Archaic Foraging Strategies and
Complexity in Northwest Alabama." Paper presented at the 67th
Annual Meeting of the Southeastern Archaeological Conference,
Lexington, KY.

Carr, Phillip J., and Andrew P. Bradbury

2000   Contemporary Lithic Analysis and Southeastern Archaeology. *Southeastern Archaeology* 19(2):120–134.

Carr, Phillip J., Andrew P. Bradbury, and Sarah E. Price

2012   Lithic Studies in the Southeast: Retrospective and Future Potential. In
*Contemporary Lithic Analysis in the Southeast: Problems, Solutions, and
Interpretations*, edited by Phillip J. Carr, Andrew P. Bradbury, and Sarah
E. Price, pp. 1–12. University of Alabama Press, Tuscaloosa.

Cashdan, Elizabeth

1983   Territoriality among Human Foragers: Ecological Models and an
Application to Four Bushman Groups. *Current Anthropology* 24(1):
47–66.

Chapman, Jefferson

1975   *The Rose Island Site and the Bifurcate Point Tradition*. Report of Investigations No. 14. Department of Anthropology, University of Tennessee, Knoxville.

1976   The Archaic Period in the Lower Little Tennessee River Valley: The Radiocarbon Dates. *Tennessee Anthropologist* 1(1):1–12.

1988   *The Archaeological Collections at the Frank H. McClung Museum*. McClung Museum Occasional Paper No. 7. Frank H. McClung Museum, University of Tennessee, Knoxville.

Chapman, Jefferson, and Andrea Brewer Shea

1981   The Archaeobotanical Record: Early Archaic Period to Contact in the Lower Little Tennessee River Valley. *Tennessee Anthropologist* 6(1):61–84.

Charnov, Eric L.

1976   Optimal Foraging: The Marginal Value Theorem. *Theoretical Population Biology* 9(2):129–136.

Cheshier, Joseph, and Robert L. Kelly

2006   Projectile Point Shape and Durability: The Effect of Thickness:Length. *American Antiquity* 71(2):353–363.

Childe, V. Gordon

1936   *Man Makes Himself*. Watts, London.

Claassen, Cheryl P.

1996   A Consideration of the Social Organization of the Shell Mound Archaic. In *Archaeology of the Mid-Holocene Southeast*, edited by Kenneth E. Sassaman and David G. Anderson, pp. 235–258. University Press of Florida, Gainesville.

Clarke, David L.

1968   *Analytical Archaeology*. Methuen, London.

Clarkson, Chris

2002   Holocene Scraper Reduction, Technological Organization and Landuse at Ingaladdi Rockshelter, Northern Australia. *Archaeology in Oceania* 37:79–86.

Clutton-Brock, Juliet

1999   *A Natural History of Domesticated Mammals*. Cambridge University Press, Cambridge.

Codding, Brian F., and Douglas W. Bird

2015   Behavioral Ecology and the Future of Archaeological Science. *Journal of Archaeological Science* 56:9–20.

Codding, Brian F., and Terry L. Jones

2013   Environmental Productivity Predicts Migration, Demographic, and

Linguistic Patterns in Prehistoric California. *Proceedings of the National Academy of Sciences* 110(36):14569–14573.

Coe, Joffre

1964 The Formative Cultures of the Carolina Piedmont. *Transactions of the American Philosophical Society* 54 (5).

Cohen, Mark N., and George J. Armelagos

1984 *Paleopathology at the Origins of Agriculture.* Academic Press, Orlando, Florida.

COHMAP Members

1988 Climatic Changes of the Last 18,000 Years: Observations and Model Simulations. *Science* 241(4869):1043–1052.

Collins, Michael B.

1999 *Clovis Blade Technology.* University of Texas Press, Austin.

Cowan, C. Wesley

1985 Understanding the Evolution of Plant Husbandry in Eastern North America: Lessons from Botany, Ethnography and Archaeology. In *Prehistoric Food Production in North America*, edited by Richard I. Ford, pp. 205–243. Museum of Anthropology, University of Michigan, Ann Arbor.

Cramér, Harald

1946 *Mathematical Methods of Statistics.* Princeton University Press, Princeton.

Crane, H. R., and James B. Griffin

1968 University of Michigan Radiocarbon Dates XII. *Radiocarbon* 10(1): 61–114.

1972 University of Michigan Radiocarbon Dates XIV. *Radiocarbon* 14(1): 155–194.

Crites, Gary D.

1987 Middle and Late Holocene Ethnobotany of the Hayes Site (40ML139): Evidence from Unit 990N918E. *Midcontinental Journal of Archaeology* 12:3–32.

1993 Domesticated Sunflower in Fifth Millennium B.P. Temporal Context: New Evidence from Middle Tennessee. *American Antiquity* 58(1): 146–148.

Crumley, Carole L.

1994 Historical Ecology: A Multidimensional Ecological Orientation. In *Historical Ecology: Cultural Knowledge and Changing Landscapes*, edited by Carole L. Crumley, pp. 1–16. School of American Research Press, Santa Fe.

Daniel, I. Randolf, Jr.

2001   Stone Raw Material Variability and Early Archaic Settlement in the
       Southeastern United States. *American Antiquity* 66(2):237–265.

Daniel, I. Randolf, Jr. and Albert C. Goodyear

2006   An Update on the North Carolina Fluted-Point Survey. *Current
       Research in the Pleistocene* 23:99–101.

DeJarnette, David L., Edward B. Kurjack, and James W. Cambron

1962   Stanfield-Worley Bluff Shelter Excavations. *Journal of Alabama Archae-
       ology* 8(1–2):1–124.

Delcourt, Hazel R.

1979   Late Quaternary Vegetation History of the Eastern Highland Rim and
       Adjacent Cumberland Plateau of Tennessee. *Ecological Monographs*
       49(3):225–280.

Delcourt, Hazel R., Paul A. Delcourt, and E. C. Spiker

1983   A 12,000-Year Record of Forest History from Cahaba Pond, St. Clair
       County, Alabama. *Ecology* 64:874–887.

Delcourt, Paul A., and Hazel R. Delcourt

1983   Late-Quaternary Vegetational Dynamics and Community Stability
       Reconsidered. *Quaternary Research* 19(2):265–271.

1985   Quaternary Palynology and Vegetational History of the Southeastern
       United States. In *Pollen Records of Late-Quaternary North American
       Sediments*, edited by Vaughn M. Bryant, Jr. and Richard G. Holloway,
       pp. 1–37. American Association of Stratigraphic Palynologists Founda-
       tion, Dallas.

2004   *Prehistoric Native Americans and Ecological Change: Human Ecosystems
       in Eastern North America in the Pleistocene.* Cambridge University
       Press, Cambridge.

Dent, Richard J.

2007   Seed Collecting and Fishing at the Shawnee Minisink Paleoindian
       Site: Everyday Life in the Late Pleistocene. In *Foragers of the Terminal
       Pleistocene*, edited by Renee B. Walker and Boyce N. Driskell, pp. 116–
       131. University of Alabama Press, Tuscaloosa.

Deter-Wolf, Aaron

2004   The Ensworth School Site (40DV184): A Middle Archaic Benton
       Occupation along the Harpeth River Drainage in Middle Tennessee.
       Tennessee Archaeology 1(1):18–35.

Deter-Wolf, Aaron, Jesse W. Tune, and John B. Broster

2011   Excavations and Dating of Late Pleistocene and Paleoindian Deposits at
       the Coats-Hines Site, Williamson County, Tennessee. *Tennessee Archae-
       ology* 5(2):142–156.

Diamond, Jared
2002   Evolution, Consequences, and Future of Plant and Animal Domestica-
       tion. *Nature* 418:700–707.
Dunnell, Robert C.
1990   The Role of the Southeast in American Archaeology. *Southeastern
       Archaeology* 9:11–22.
Dunnell, Robert C., and William S. Dancey
1983   The Siteless Survey: A Regional Scale Data Collection Strategy. In
       Advances in Archaeological Method and Theory, edited by Michael B.
       Schiffer, pp. 267–287. Academic Press, New York.
Dye, David H.
2013   Trouble in the Glen: The Battle over Kentucky Lake Archaeology.
       In *Shovel Ready: Archaeology and Roosevelt's New Deal for America*,
       edited by Bernard K. Means, pp. 129–146. University of Alabama Press,
       Tuscaloosa.
Dyson-Hudson, Rada, and Eric Alden Smith
1978   Human Territoriality: An Ecological Reassessment. *American Anthro-
       pologist* 80(1):21–41.
Eaton, S. Boyd, Majorie Shostak, and Melvin Konner
1988   *The Paleolithic Prescription: A Program of Diet and Exercise and a
       Design for Living.* Harper & Row, New York.
Eren, Metin I., and Brian N. Andrews
2013   Were Bifaces Used as Mobile Cores by Clovis Foragers in the North
       American Lower Great Lakes Region? An Archaeological Test of
       Experimentally Derived Quantitative Predictions. *American Antiquity*
       78(1):166–180.
Eren, Metin I., and Mary E. Prendergast
2008   Comparing and Synthesizing Lithic Reduction Indices. In *Lithic Tech-
       nology: Measures of Production, Use, and Curation*, edited by William
       Andrefsky, Jr., pp. 49–85. Cambridge University Press, Cambridge.
Faulkner, Charles H., and Major C. R. McCollough
1973   *Introductory Report of the Normandy Reservoir Salvage Project: Environ-
       mental Setting, Typology, and Survey*, Vol. 1, Report of Investigations
       No. 11. Department of Anthropology, University of Tennessee, Knoxville.
Fenneman, Nevin M.
1938   *Physiography of the Eastern United States.* McGraw-Hill, New York.
Ferring, C. Reid
2012   "The 'Long' Clovis Chronology: Evidence from the Aubrey and
       Friedkin Sites, Texas." Paper presented at the 77th Annual Meeting of
       the Society for American Archaeology, Memphis, TN.

Fiedel, Stuart, and Gary Haynes
2004   A Premature Burial: Comments on Grayson and Meltzer's "Requiem for Overkill." *Journal of Archaeological Science* 31:121–131.

Figurska, Małgorzata, Maciej Stańczyk, and Kamil Kulesza
2008   Humans Cannot Consciously Generate Random Numbers Sequences: Polemic Study. *Medical Hypotheses* 70(1):182–185.

Fisher, Ronald
1918   The Correlation between Relatives on the Supposition Of Mendelian Inheritance. *Philosophical Transactions of the Royal Society of Edinburgh* 52:399–433.

Flannery, Kent V.
1969   Origins and Ecological Effects of Early Domestication in Iran and the Near East. In *The Domestication and Exploitation of Plants and Animals*, edited by Peter J. Ucko and G. W. Dimbleby, pp. 73–100. Aldine, Chicago.

Ford, James A., and Clarence H. Webb
1956   *Poverty Point, a Late Archaic Site in Louisiana.* Anthropological Papers Vol. 46, Pt. 1. American Museum of Natural History, New York.

Franklin, Jay D.
2002   "The Prehistory of Fentress County, Tennessee: An Archaeological Survey." Ph.D. dissertation, Department of Anthropology, University of Tennessee, Knoxville.

Fretwell, Steven Dewitt, and Henry. L. Lucas, Jr.
1969   On Territorial Behavior and Other Factors Influencing Habitat Distribution in Birds. *Acta Biotheoretica* 19:16–36.

Fritz, Gayle J.
1990   Multiple Pathways to Farming in Precontact Eastern North America. *Journal of World Prehistory* 4:387–435.
1997   A Three-Thousand-Year-Old Cache of Crop Seeds from Marble Bluff, Arkansas. In *People, Plants, and Landscapes: Studies in Paleoethnobotany*, edited by Kristen J. Gremillion, pp. 42–62. University of Alabama Press, Tuscaloosa.

Gardner, Paul S.
1997   The Ecological Structure and Behavioral Implications of Mast Exploitation Strategies. In *People, Plants, and Landscapes: Studies in Paleoethnobotany*, edited by Kristen J. Gremillion, pp. 161–178. University of Alabama Press, Tuscaloosa.

Germonpré, Mietje, Martina Lázničková-Galetová, and Mikhail Sablin
2012   Palaeolithic Dog Skulls at the Gravettian Předmostí Site, the Czech Republic. *Journal of Archaeological Science* 39(1):184–202.

Gingerich, Joseph A. M.

2007 "Shawnee-Minisink Revisited: New Excavations of the Paleoindian Occupation." Master's thesis, Department of Anthropology, University of Wyoming, Laramie.

2011 Down to Seeds and Stones: A New Look at the Subsistence Remains from Shawnee-Minisink. *American Antiquity* 76(1):127–144.

Goodyear, Albert C.

1974 *The Brand Site: A Techno-Functional Study of a Dalton Site in Northeast Arkansas.* Research Series No. 7. Arkansas Archaeological Survey, Fayetteville.

1979 *A Hypothesis for the Use of Cryptocrystalline Raw Materials among Paleo-Indian Groups of North America.* South Carolina Institute of Archaeology and Anthropology, University of South Carolina, Columbia.

1982 The Chronological Position of the Dalton Horizon in the Southeastern United States. *American Antiquity* 47(2):382–395.

1989 A Hypothesis for the Use of Cryptocrystalline Raw Materials among Paleoindian Groups of North America. In *Eastern Paleoindian Lithic Resource Use*, edited by Chris J. Ellis and Jonathan C. Lothrop, pp. 1–9. Westview, Boulder, CO.

1999 The Early Holocene Occupation of the Southeastern United States: A Geoarchaeological Summary. In *Ice Age Peoples of North America*, edited by Robson Bonnichsen and Karen L. Turnmire, pp. 432–481. Oregon State University Press, Corvallis.

2006 Recognizing the Redstone Fluted Point in the South Carolina Paleoindian Point Database. *Current Research in the Pleistocene* 23:112–114.

Gould, Richard A.

1977 *Puntutjarpa Rockshelter and the Australian Desert Culture.* Anthropological Papers Vol. 54, Pt. 1. American Museum of Natural History, New York.

1978 Beyond Analogy in Ethnoarchaeology. In *Explorations in Ethnoarchaeology*, edited by Richard A. Gould, pp. 249–293. University of New Mexico Press, Albuquerque.

1980 *Living Archaeology.* Cambridge University Press, New York.

1996 Faunal Reduction at Puntutjarpa Rockshelter, Warburton Ranges, Western Australia. *Archaeology in Oceania* 31(2):72–86.

Graham, Russell W., and Ernest L. Lundelius, Jr.

2010 FAUNMAP II: New Data for North America with a Temporal Extension for the Blancan, Irvingtonian, and Early Rancholabrean. FAUNMAP II Database, version 1.0.

Graham, R. W., C. V. Haynes, D. L. Johnson, and M. Kay
1981   Kimmswick: A Clovis-Mastodon Association in Eastern Missouri.
        Science 213:1115–1117.
Gramly, Richard M.
2009   *Origin and Evolution of the Cumberland Palaeo-American Tradition.*
        American Society for Amateur Archaeology, North Andover, MA.
Gramly, Richard M., and Richard E. Funk
1991   Olive Branch: A Large Dalton and Pre-Dalton Encampment at Thebes
        Gap, Alexander County, Illinois. In *The Archaic Period in the Mid-
        South: Proceedings of the 1989 Mid-South Archaeological Conference,
        Memphis, Tennessee—July 15, 1989,* edited by Charles H. McNutt,
        pp. 23–34. Archaeological Report No. 24, Mississippi Department of
        Archives and History, Jackson.
Grayson, Donald K.
1983   The Paleontology of Gatecliff Shelter. In *The Archaeology of Monitor
        Valley: 2. Gatecliff Shelter,* by David Hurst Thomas, pp. 99–125. Anthro-
        pological Papers Vol. 59, Pt. 1. American Museum of Natural History,
        New York.
Grayson, Donald K., and David J. Meltzer
2003   A Requiem for North American Overkill. *Journal of Archaeological
        Science* 30:585–593.
2004   North American Overkill Continued? *Journal of Archaeological Science*
        31:133–136.
2015   Revisiting Paleoindian Exploitation of Extinct North American Mam-
        mals. *Journal of Archaeological Science* 56:177–193.
Gremillion, Kristen J.
1996   The Paleoethnobotanical Record for the Mid-Holocene Southeast. In
        Archaeology of the Mid-Holocene Southeast, edited by Kenneth E. Sassa-
        man and David G. Anderson, pp. 99–114. University Press of Florida,
        Gainesville.
2002   The Development and Dispersal of Agricultural Systems in the Wood-
        land Period Southeast. In *The Woodland Southeast,* edited by David G.
        Anderson and Robert C. Mainfort, Jr., pp. 483–501. University of Ala-
        bama Press, Tuscaloosa.
2004   Seed Processing and the Origins of Food Production in Eastern North
        America. *American Antiquity* 69(2):215–233.
Gremillion, Kristen J., Loukas Barton, and Dolores R. Piperno
2014a   Particularism and the Retreat from Theory in the Archaeology of
        Agricultural Origins. *Proceedings of the National Academy of Sciences*
        111(17):6171–6177.
2014b   Reply to Smith: On Distinguishing between Models, Hypotheses, and

Theoretical Frameworks. *Proceedings of the National Academy of Sciences* 111(28):E2830.

2014c   Reply to Zeder: Maintaining a Diverse Scientific Toolkit Is Not an Act of Faith. *Proceedings of the National Academy of Sciences* 111(28):E2828.

Gremillion, Kristen J., and Dolores R. Piperno

2009   Human Behavioral Ecology, Phenotypic (Developmental) Plasticity, and Agricultural Origins: Insights from the Emerging Evolutionary Synthesis. *Current Anthropology* 50(5):615–619.

Gremillion, Kristen J., Jason Windingstad, and Sarah C. Sherwood

2008   Forest Opening, Habitat Use, and Food Production on the Cumberland Plateau, Kentucky: Adaptive Flexibility in Marginal Settings. *American Antiquity* 73(3):387–411.

Griffin, James B.

1952   Culture Periods in Eastern United States Archeology. In *Archeology of the Eastern United States*, edited by James B. Griffin, pp. 352–364. University of Chicago Press, Chicago.

1967   Eastern North American Archaeology: A Summary. *Science* 156(3772): 175–191.

Hamilton, Marcus J., and Briggs Buchanan

2009   Spatial Gradients in Clovis-Age Radiocarbon Dates across North America Suggest Rapid Colonization from the North. *Proceedings of the National Academy of Sciences* 104(40):15625–15630.

Haven, Samuel F.

1856   *Archaeology of the United States, Or Sketches, Historical and Biographical, of the Progress of Information and Opinion Respecting Vestiges of Antiquity in the United States.* Smithsonian Institution, Washington, DC.

Hayden, Brian

1992   Models of Domestication. In *Transitions to Agriculture in Prehistory*, edited by Anne Birgitte Gebauer and T. Douglas Price, pp. 11–199. Monographs in World Archaeology No. 4. Prehistory Press, Madison, WI.

1995   A New Overview of Domestication. In *Last Hunters—First Farmers: New Perspectives on the Prehistoric Transition to Agriculture*, edited by T. Douglas Price and Anne Birgitte Gebauer, pp. 273–299. School of American Research Press, Sante Fe.

Haynes, Gary, David G. Anderson, C. Reid Ferring, Stuart J. Fiedel, Donald K. Grayson, C. Vance Haynes, Jr., Vance T. Holliday, Bruce B. Huckell, Marcel Kornfeld, David J. Meltzer, Julie Morrow, Todd Surovell, Nicole M. Waguespack, Peter Wigand, and Robert M. Yohe II

2007   Comment on "Redefining the Age of Clovis: Implications for the Peopling of the Americas." *Science* 317(5836):320b.

Hemmings, C. Andrew

1998 Probable Association of Paleoindian Artifacts and Mastodon Remains from Sloth Hole, Aucilla River, North Florida. *Current Research in the Pleistocene* 15:16–18.

Herrmann, Nicholas P.

2002 "Biological Affinities of Archaic Period Populations from West-Central Kentucky and Tennessee." Ph.D. dissertation, Department of Anthropology, University of Tennessee, Knoxville.

Higgins, Katherine French

1982 "The Ledbetter Landing Site: A Study in Late Archaic Mortuary Patterning." Master's thesis, Department of Anthropology, University of Tennessee, Knoxville.

Hiscock, Peter, and Chris Clarkson

2005 Experimental Evaluation of Kuhn's Geometric Index of Reduction and the Flat-Flake Problem. *Journal of Archaeological Science* 32:1015–1022.

Hoffman, C. Marshall

1985 Projectile Point Maintenance and Typology: Assessment with Factor Analysis and Canonical Correlation. In *For Concordance in Archaeological Analysis: Bridging Data Structure, Quantitative Technique, and Theory*, edited by Christopher Carr, pp. 566–612. Westport, Kansas City, MO.

Hollenbach, Kandice D.

2007 Gathering in the Late Paleoindian: Archaeobotanical Remains from Dust Cave, Alabama. In *Foragers of the Terminal Pleistocene in North America*, edited by Renee B. Walker and Boyce N. Driskell, pp. 132–147. University of Nebraska Press, Lincoln.

2009 *Foraging in the Tennessee Valley, 12,500 to 8,000 Years Ago.* University of Alabama Press, Tuscaloosa.

Hollenbach, Kandace D., and Renee B. Walker

2010 Investigations of Paleoethnobotanical and Zooarchaeological Data from Dust Cave, Alabama. In *Integrating Zooarchaeology and Paleoethnobotany: A Consideration of Issues, Methods, and Cases*, edited by Amber M. VanDerwarker and Tanya M. Peres, pp. 227–244. Springer, New York.

Holliday, Vance T., and D. Shane Miller

2013 The Clovis Landscape. In *Paleoamerican Odyssey*, edited by Kelly E. Graf, Caroline V. Ketron, and Michael R. Waters, pp. 221–245. Texas A&M University Press, College Station.

Holmes, William Henry

1890 A Quarry Workshop of the Flaked-Stone Implement Makers in the District of Columbia. *American Anthropologist* 3(1):1–26.

Howard, Calvin D.

1990 The Clovis Point: Characteristics and Type Description. *Plains Anthropologist* 35:255–262.

Hunzicker, David A.

2008 Folsom Projectile Point Technology: An Experiment in Design, Effectiveness, and Efficiency. *Plains Anthropologist* 53(207):291–311.

Ingbar, Eric E.

1994 Lithic Material Selection and Technological Organization. In *The Organization Of North American Prehistoric Chipped Stone Tool Technologies*, edited by Phillip J. Carr, pp. 45–56. Archaeological Series No. 7. International Monographs in Prehistory, Ann Arbor.

Ingmanson, John E., and John W. Griffin

1974 Material Culture. In *Investigations in Russell Cave*, edited by John W. Griffin. Publications in Archaeology No. 13. National Park Service, US Department of the Interior, Washington, DC.

Iovita, Radu, Holger Schönekess, Sabine Gaudzinski-Windheuser, and Frank Jäger

2014 Projectile Impact Fractures and Launching Mechanisms: Results of a Controlled Ballistic Experiment Using Replica Levallois Points. *Journal of Archaeological Science* 48:73–83.

Jazwa, Christopher S., Douglas J. Kennett, and Bruce Winterhalder

2016 A Test of Ideal Free Distribution Predictions Using Targeted Survey and Excavation on California's Northern Channel Islands. *Journal of Archaeological Method and Theory* 23(4):1242–1284.

Jefferies, Richard W.

1996 The Emergence of Long-Distance Exchange Networks in the Southeastern United States. In *Archaeology of the Mid-Holocene Southeast*, edited by Kenneth E. Sassaman and David G. Anderson, pp. 222–234. University Press of Florida, Gainesville.

2004 Regional-Scale Interaction Networks and the Emergence of Cultural Complexity along the Northern Margins of the Southeast. In *Signs of Power: The Rise of Cultural Complexity in the Southeast*, edited by Jon L. Gibson and Philip J. Carr, pp. 71–85. University of Alabama Press, Tuscaloosa.

Jennings, Thomas A.

2008 San Patrice: An Example of Late Paleoindian Adaptive Versatility in South-Central North America. *American Antiquity* 73(3):539–559.

Jochim, Michael A.

1991 Archaeology as Long-Term Ethnography. *American Anthropologist* 93(2):308–321.

Jolley, Robert L.
1980   *An Archaeological Survey of the Lower Duck and Middle Cumber-
       land Rivers in Middle Tennessee.* Tennessee Division of Archaeology,
       Nashville.

Jones, Volney
1936   The Vegetal Remains of Newt Kash Hollow Shelter. In *Rock Shelters in
       Menifee County, Kentucky*, edited by William S. Webb and W. D. Funk-
       houser, pp. 147–165. Reports in Archaeology and Anthropology Vol. 3,
       No. 4. University of Kentucky, Lexington.

Justice, Noel D.
1995   *Stone Age Spear and Arrow Points of the Midcontinental and Eastern
       United States: A Modern Survey and Reference.* Indiana University Press,
       Bloomington, IN.

Keeley, Lawrence H.
1982   Hafting and Retooling: Effects on the Archaeological Record. *American
       Antiquity* 47(4):798–809.

Kelly, Robert L.
1988   The Three Sides of a Biface. *American Antiquity* 53(4):717–734.
1992   Mobility/Sedentism: Concepts, Archaeological Measures, and Effects.
       Annual Review of Anthropology 21:43–66.
1995   *The Foraging Spectrum: Diversity in Hunter-Gatherer Lifeways.* Smith-
       sonian Institution Press, Washington, DC.
2013   *The Lifeways of Hunter-Gatherers: The Foraging Spectrum.* Cambridge
       University Press, Cambridge.

Kelly Robert L., and Lawrence C. Todd
1988   Coming into the Country: Early Paleoindian Hunting and Mobility.
       American Antiquity 53(2):231–244.

Kelly, Robert L., Todd A. Surovell, Bryan N. Shuman, and Geoffrey M. Smith
2013   A Continuous Climatic Impact on Holocene Human Population in
       the Rocky Mountains. *Proceedings of the National Academy of Sciences*
       110:443–447.

Kennett, Douglas J., Atholl Anderson, and Bruce Winterhalder
2006   The Ideal Free Distribution, Food Production, and the Colonization of
       Oceania. In *Behavioral Ecology and the Transition to Agriculture*, edited
       by Douglas J. Kennett and Bruce Winterhalder, pp. 265–288. University
       of California Press, Berkeley.

Kennett, Douglas J., and Bruce Winterhalder (editors)
2006   *Behavioral Ecology and the Transition to Agriculture.* University of Cali-
       fornia Press, Berkeley.

Kennett, Douglas J., and Bruce Winterhalder
2008   Demographic Expansion, Despotism, and the Colonisation of East and
       South Polynesia. In *Islands of Inquiry: Colonisation, Seafaring and the*

Archaeology of Maritime Landscapes, edited by Geoffrey Clark, Foss Leach, and Sue O'Connor, pp. 87–96. Australia National University Press, Canberra.

Kerr, Jonathan P.

1996 *Archeological Survey of Kentucky Lake, Western Tennessee and Kentucky*. Contract Publication Series 96–14, Cultural Resource Analysts, Lexington, KY.

Kerr, Jonathan P., and Andrew P. Bradbury

1998 Paleo-Indian and Archaic Settlement at Kentucky Lake. *Tennessee Anthropologist* 23(1–2):1–20.

Kilby, J. David

2008 "An Investigation of Clovis Caches: Content, Function, and Technological Organization." Ph.D. dissertation, Department of Anthropology, University of New Mexico, Albuquerque.

Kimball, Larry R.

1981 "An Analysis of Residential Camp Site Structure for Two Early Archaic Assemblages from Rose Island (40MR44)." Master's thesis. University of Tennessee, Knoxville.

1996 Early Archaic Settlement and Technology: Lessons from Tellico. In *The Paleoindian and Early Archaic Southeast*, edited by David G. Anderson and Kenneth E. Sassaman, pp. 149–186. University of Alabama Press, Tuscaloosa.

Klippel, Walter E., and Paul W. Parmalee

1982 *Paleontology of Cheek Bend Cave, Maury County, Tennessee: Phase II Report*. Department of Anthropology, University of Tennessee, Knoxville.

Kramer, Clyde Young

1956 Extension of Multiple Range Tests to Group Means with Unequal Numbers of Replications. *Biometrics* 12:307–310.

Kuhn, Steven L.

1990 A Geometric Index of Reduction for Unifacial Stone Tools. *Journal of Archaeological Science* 17:583–593.

1994 A Formal Approach to the Design and Assembly of Mobile Toolkits. *American Antiquity* 59(3):426–442.

1995 *Mousterian Lithic Technology*. Princeton University Press, Princeton.

2004 Upper Paleolithic Raw Material Economies at Üçağızlı Cave, Turkey. *Journal of Anthropological Archaeology* 23:431–448.

Kuhn, Steven L., and D. Shane Miller

2015 Artifacts as Patches: The Marginal Value Theorem and Stone Tool Life Histories. In *Lithic Technological Systems and Evolutionary Theory*, edited by Nathan Goodale and William Andrefsky, Jr., pp. 172–197. Cambridge University Press, Cambridge.

Larsen, Clark Spencer
2002 Post-Pleistocene Human Evolution: Bioarchaeology of the Agricultural
      Transition. In *Human Diet: Its Origin and Evolution*, edited by Peter S.
      Ungar and Mark F. Teaford, pp. 19–35. Greenwood, Westport, CT.
Leach, Elizabeth K., and Michael T. Jackson
1987 Geomorphic History of the Lower Cumberland and Tennessee Valleys
      and Implications for Regional Archaeology. *Southeastern Archaeology*
      6(2):100–107.
Leigh, David S.
2008 Late Quaternary Climates and River Channels of the Atlantic Coastal
      Plain, Southeastern USA. *Geomorphology* 101:90–108.
Leonard, Jennifer A., Robert K. Wayne, Jane Wheeler, Raúl Valadez, Sonia
Guillén, and Carles Vilà
2002 Ancient DNA Evidence for Old World Origin of New World Dogs.
      Science 298(5598):1613–1616.
Lewis, Thomas M. N.
1954a A Suggested Basis for Paleo-Indian Chronology in Tennessee and the
      Eastern United States. *Southern Indian Studies* 6:11–13.
1954b The Cumberland Point. *Bulletin of the Oklahoma Anthropological
      Society* 11:7–8.
Lewis, Thomas M. N., and Madeline Kneberg
1958 The Nuckolls Site: A Possible Dalton-Meserve Chipped Stone Complex
      in the Kentucky Lake Area. *Tennessee Archaeologist* 14(2):60–79.
1959 The Archaic Culture in the Middle South. *American Antiquity* 25(2):
      161–183.
Lewis, Thomas M. N., and Madeline Kneberg Lewis
1961 *Eva, an Archaic Site*. University of Tennessee Study in Anthropology,
      University of Tennessee Press, Knoxville.
Libby, Willard F.
1952 *Radiocarbon Dating*. University of Chicago Press, Chicago.
Lorenz, Edward N.
1963 Deterministic Nonperiodic Flow. *Journal of the Atmospheric Sciences*
      20(2):130–141.
Lyman, R. Lee
1994 *Vertebrate Taphonomy*. Cambridge Manuals in Archaeology.
      Cambridge University Press, Cambridge.
MacArthur, Robert H., and Eric R. Pianka
1966 On Optimal Use of a Patchy Environment. *American Naturalist*
      100(196):603–609.
Magennis, Ann L.
1977 "Middle and Late Archaic Mortuary Patterning: An Example from the

Western Tennessee Valley." Master's thesis, Department of Anthropology, University of Tennessee, Knoxville.

Malthus, Thomas
1798   *An Essay on the Principle of Population, as It Affects the Future Improvement of Society. With Remarks on the Speculations of Mr. Godwin,
M. Condorcet and Other Writers.* J. Johnson, London.

Mann, Charles C.
2006   *1491: New Revelations of the Americas before Columbus.* Vintage, New
York.

Marquardt, William H., and Patty Jo Watson
2005   *Archaeology of the Middle Green River Region, Kentucky.* University
Press of Florida, Gainesville.

McClure, Sarah B., Michael A. Joachim, and C. Michael Barton
2006   Human Behavioral Ecology, Domestic Animals, and Land Use during
the Transition to Agriculture in Valencia, Eastern Spain. In *Behavioral
Ecology and the Transition to Agriculture*, edited by Douglas J. Kennett
and Bruce Winterhalder. University of California Press, Berkeley.

McGovern, Thomas H.
1994   Management for Extinction in Norse Greenland. In *Historical
Ecology: Cultural Knowledge and Changing Landscapes*, edited by
Carole L. Crumley, pp. 127–154. School of American Research Press,
Sante Fe.

McGuire, Kelly R., and William R. Hildebrandt
2005   Re-Thinking Great Basin Foragers: Prestige Hunting and Costly Signaling during the Middle Archaic Period. *American Antiquity* 70(4):
695–712.

McNutt, Charles H.
2008   The Benton Phenomenon and Middle Archaic Chronology in Adjacent
Portions of Tennessee, Mississippi, and Alabama. *Southeastern Archaeology* 27(1):45–60.

McNutt, Charles H., John B. Broster, and Mark R. Norton
2008   A Surface Collection from the Kirk Point Site (40HS174), Humphreys
County, Tennessee. *Tennessee Archaeology* 3(1):25–75.

Means, Bernard K.
2013   "Alphabet Soup" and American Archaeology. In *Shovel Ready: Archaeology and Roosevelt's New Deal for America*, edited by Bernard K. Means,
pp. 1–20. University of Alabama Press, Tuscaloosa.

Meeks, Scott C.
2000   *The Use and Function of Late Middle Archaic Projectile Points in the
Midsouth.* Report of Investigations No. 77. Office of Archaeological
Services, University of Alabama Museums, Moundville.

Meltzer, David J.

1988   Late Pleistocene Human Adaptations in Eastern North America. *Journal of World Prehistory* 2:1–52.

2006   *Folsom: New Archaeological Investigations of a Classic Paleoindian Bison Kill.* University of California Press, Berkeley.

2009   *First Peoples in a New World: Colonizing Ice Age America.* University of California Press, Berkeley.

Meltzer, David J., and Vance T. Holliday

2010   Would North American Paleoindians Have Noticed Younger Dryas Age Climate Changes? *Journal of World Prehistory* 23:1–41.

Meltzer, David J., and Bruce D. Smith

1986   Paleoindian and Early Archaic Subsistence Strategies in Eastern North America. In *Foraging, Collecting, Harvesting: Archaic Period Subsistence and Settlement in the Eastern Woodlands*, edited by Sarah W. Nesius, pp. 3–31. Occasional Paper No. 6, Center for Archaeological Investigations, Southern Illinois University, Carbondale.

Mensforth, Robert P.

2007   Human Trophy Taking in Eastern North America during the Archaic Period: The Relationship to Warfare and Social Complexity. In *The Taking and Displaying of Human Body Parts as Trophies by Amerindians*, edited by Richard J. Chacon and David H. Dye, pp. 222–277. Springer, New York.

Metcalfe, Duncan, and K. Renee Barlow

1992   A Model for Exploring the Optimal Trade-off between Field Processing and Transport. *American Anthropologist* 94(2):340–356.

Mill, John Stuart

1844   Essay V: On the Definition of Political Economy; and on the Method of Investigation Proper to It. In *Essays on Some Unsettled Questions of Political Economy*, edited by John Stuart Mill, pp. 54–74. John W. Parker, London.

Miller, D. Shane, and Joseph A. M. Gingerich

2013a Paleoindian Chronology and the Eastern Fluted Point Tradition. In *In the Eastern Fluted Point Tradition*, edited by Joseph A. M. Gingerich, pp. 9–37. University of Utah Press, Salt Lake City.

2013b Regional Variation in the Terminal Pleistocene and Early Holocene Radiocarbon Record of Eastern North America. *Quaternary Research* 79:175–188.

Miller, D. Shane, Vance T. Holliday, and Jordon Bright

2013   Clovis across the Continent. In *Paleoamerican Odyssey*, edited by Kelly E. Graf, Caroline V. Ketron, and Michael R. Waters, pp. 207–220.

Center for the Study of the First Americans, Texas A&M University Press, College Station.

Miller, D. Shane, and Ashley M. Smallwood

2012 Beyond Stages: Modeling Clovis Biface Production at the Topper Site, South Carolina. In *Contemporary Lithic Analysis in the Southeast: Problems, Solutions, and Interpretations*, edited by Philip J. Carr, Andrew P. Bradbury, and Sarah E. Price, pp. 28–41. University of Alabama Press, Tuscaloosa.

Mills, H. H., and Paul A. Delcourt

1991 Quaternary Geology of the Appalachian Highlands and Interior Low Plateaus. In *Quaternary Nonglacial Geology: Conterminous United States*, edited by Roger B. Morrison, pp. 611–628. Geological Society of America, Boulder, CO.

Milner, George R.

1999 Warfare in Prehistoric and Early Historic Eastern North America. *Journal of Archaeological Research* 7(2):105–151.

Milton, Katharine

2002 Hunter-Gatherer Diets: Wild Foods Signal Relief from Diseases of Affluence. In *Human Diet: Its Origins and Evolution*, edited by Peter S. Ungar and Mark F. Teaford, pp. 111–122. Greenwood, Westport, CT.

Mlodinow, Leonard

2008 *The Drunkard's Walk: How Randomness Rules Our Lives*. Vintage, New York.

Moore, Christopher R., and Victoria G. Dekle

2010 Hickory Nuts, Bulk Processing, and the Advent of Early Horticultural Economies in Eastern North America. *World Archaeology* 42(4): 595–608.

Morlan, Richard

2005 *The Canadian Archaeological Radiocarbon Database*. Canadian Museum of Civilization, Ottawa.

Morrow, Juliet E.

1995 Clovis Projectile Point Manufacture: A Perspective from the Ready/Lincoln Hills Site, 11JY46, Jersey County, Illinois. *Midcontinental Journal of Archaeology* 20(2):167–191.

Morse, Dan F.

1967 "The Robinson Site and Shell Mound Archaic Culture in the Middle South." Ph.D. dissertation, Department of Anthropology, University of Michigan, Ann Arbor.

1973 Dalton Culture in Northeastern Arkansas. *Florida Anthropologist* 26(1): 23–38.

Nadel, Dani, and Ella Werker

1999   The Oldest Ever Brush Hut Plant Remains from Ohalo II, Jordan Valley,
       Israel (19,000 BP). *Antiquity* 73:755–764.

Nelson, Margaret C.

1991   The Study of Technological Organization. *Archaeological Method and
       Theory* 3:57–100.

Norton, Mark R., and John B. Broster

1992   40HS200: The Nuckolls Extension Site. *Tennessee Anthropologist* 17:
       13–32.

O'Connell, James F.

1995   Ethnoarchaeology Needs a General Theory of Behavior. *Journal of
       Archaeological Research* 3(3):205–255.

O'Connell, James F., and Jim Allen

2012   The Restaurant at the End of the Universe: Modelling the Colonisation
       of Sahul. *Australian Archaeology* 74:5–17.

Odell, George H., and Frank Cowan

1986   Experiments with Spears and Arrows on Animal Targets. *Journal of
       Field Archaeology* 13:195–212.

Odling-Smee, John, Douglas H. Erwin, Eric P. Palkovacs, Marcus W. Feldman,
and Kevin N. Laland

2013   Niche Construction Theory: A Practical Guide for Ecologists. *Quarterly
       Review of Biology* 88(1):3–28.

Osborne, Douglas

1942   "The Big Sandy Site, Henry County, Tennessee." Master's thesis, Depart-
       ment of Anthropology, University of New Mexico, Albuquerque.

Parish, Ryan

2011   The Application of Visible/Near-Infrared Reflectance (VNIR) Spectros-
       copy to Chert Sourcing: A Case Study from the Dover Quarry Sites,
       Tennessee. *Geoarchaeology* 26:420–439.

Parmalee, Paul W., and Walter E. Klippel

1982   Evidence of Boreal Avifauna in Middle Tennessee during the Late
       Pleistocene. *Auk* 99(2):365–368.

Peacock, Evan

1988   Benton Settlement Patterns in North-Central Mississippi. *Mississippi
       Archaeology* 23(1):12–33.

Pearson, Karl

1900   On the Criterion that a Given System of Deviations from the Probable
       in the Case of a Correlated System of Variables Is Such That It Can Be
       Reasonably Supposed to Have Arisen from Random Sampling. *Philo-
       sophical Magazine* 50(302):157–175.

1901   On Lines and Planes of Closest Fit to Systems of Points in Space. *Philosophical Magazine* 2(11):559–572.

Persky, Joseph

1995   Retrospectives: The Ethology of Homo Economicus. *Journal of Economic Perspectives* 9(2):221–231.

Plackett, R. L.

1983   Karl Pearson and the Chi-Squared Test. *International Statistical Review* 51(1):59–72.

Powell, Joseph F.

1995   "Dental Variation and Biological Affinity among Middle Holocene Human Populations in North America." Ph.D. dissertation, Department of Anthropology, Texas A&M University, College Station.

Prasciunas, Mary M.

2008   "Clovis First? An Analysis of Space, Time, and Technology." Ph.D. dissertation, Department of Anthropology, University of Wyoming, Laramie.

2011   Mapping Clovis: Projectile Points, Behavior, and Bias. *American Antiquity* 76(1):107–126.

Price, T. Douglas

2009   Ancient Farming in Eastern North America. *Proceedings of the National Academy of Sciences* 106(16):6427–6428.

Prufer, Keith M., Amy E. Thompson, Clayton R. Meredith, Brendan J. Culleton, Jillian M. Jordan, Claire E. Ebert, Bruce Winterhalder, and Douglas J. Kennett

2017   The Classic Period Maya Transition from an Ideal Free to an Ideal Despotic Settlement System at the Polity of Uxbenká. *Journal of Anthropological Archaeology* 45:53–68.

Raab, L. Mark, and Albert C. Goodyear

1984   Middle-Range Theory in Archaeology: A Critical Review of Origins and Applications. *American Antiquity* 49(2): 255–268.

Randall, Asa

2002   "Technofunctional Variation in Early Side-Notched Hafted Bifaces: A View from the Middle Tennessee River Valley in Northwest Alabama." Master's thesis, Department of Anthropology, University of Florida, Gainesville.

Redman, Charles L.

1999   *Human Impact on Ancient Environments*. University of Arizona Press, Tucson.

Redmond, Brian G., and Kenneth B. Tankersley

2005   Evidence of Early Paleoindian Bone Modification and Use at the Sheriden Cave Site (33WY252), Wyandot County, Ohio. *American Antiquity* 70(3):503–526.

Rindos, David
1984   *The Origins of Agriculture: An Evolutionary Perspective.* Academic Press,
        San Diego.
Rosenberg, Michael
1998   Cheating at Musical Chairs: Territoriality and Sedentism in an Evolu-
        tionary Context. *Current Anthropology* 39(5):653–681.
Ross, Cody T., and Bruce Winterhalder
2015   Sit-and-Wait versus Active-Search Hunting: A Behavioral Ecological
        Model of Optimal Search Mode. *Journal of Theoretical Biology* 387:
        76–87.
Sassaman, Kenneth E.
1993   *Early Pottery in the Southeast: Tradition and Innovation in Cooking
        Technology.* University of Alabama Press, Tuscaloosa.
2001   Hunter-Gatherers and Traditions of Resistance. In *The Archaeology
        of Traditions: Agency and History before and after Columbus*, edited
        by Timothy R. Pauketat, pp. 218–236. University Press of Florida,
        Gainesville.
2004   Complex Hunter-Gatherers in Evolution and History: A North
        American Perspective. *Journal of Archaeological Research* 12(3):
        227–280.
2010   *The Eastern Archaic, Historicized.* AltaMira, Lanham, MD.
Sassaman, Kenneth E., Meggan E. Blessing, and Asa R. Randall
2006   Stallings Island Revisited: New Observations on Occupational History,
        Community Patterning, and Subsistence Technology. *American Antiq-
        uity* 71(3):539–566.
Schiffer, Michael
1976   *Behavioral Archeology.* Academic Press, New York.
1983   Toward the Identification of Formation Processes. *American Antiquity*
        48(4):675–706.
1988   The Structure of Archaeological Theory. *American Antiquity* 53(3):
        461–485.
Sellards, Elias H.
1952   *Early Man in America: A Study in Prehistory.* University of Texas Press,
        Austin.
Shennan, Stephen, and Kevan Edinborough
2007   Prehistoric Population History: From the Late Glacial to the Late Neo-
        lithic in Central and Northern Europe. *Journal of Archaeological Science*
        34(8):1339–1345.
Sherwood, Sarah C., Boyce N. Driskell, Asa R. Randall, and Scott C. Meeks
2004   Chronology and Stratigraphy at Dust Cave, Alabama. *American Antiq-
        uity* 69(3):533–534.

Shott, Michael J.

1996   Stage versus Continuum in the Debris Assemblage from Production of
        a Fluted Biface. *Lithic Technology* 21:6–22.

1998   Status and Role of Formation Theory in Contemporary Archaeological
        Practice. *Journal of Archaeological Research* 6(4):299–329.

2004   Representivity of the Midwestern Paleoindian Site Sample. *North
        American Archaeologist* 25:189–212.

2015   Glass Is Heavy, Too: Testing the Field-Processing Model at the Modena
        Obsidian Quarry, Lincoln County, Southeastern Nevada. *American
        Antiquity* 80(3):548–570.

Shott, Michael J., and Jesse A. M. Ballenger

2007   Biface Reduction and the Measurement of Dalton Curation: A South-
        eastern United States Case Study. *American Antiquity* 72(1):153–175.

Sih, Andrew

1980   Optimal Foraging: Partial Consumption of Prey. *American Naturalist*
        116(2):281–290.

Smallwood, Ashley M.

2012   Clovis Technology and Settlement in the American Southeast: Using
        Biface Analysis to Evaluate Dispersal Models. *American Antiquity*
        77(4):689–713.

Smith, Bruce D.

1986   Archaeology of the Southeastern United States: From Dalton to De
        Soto, 10,500–500 BP. *Advances in World Archaeology* 5:1–92.

1987   Independent Domestication of Seed-Bearing Plants in Eastern North
        America. In *Emergent Horticultural Economies of the Eastern Wood-
        lands*, edited by William F. Keegan, pp. 3–47. Occasional Paper No. 7,
        Center for Archaeological Investigations, Southern Illinois University,
        Carbondale.

1992   The Floodplain Weed Theory of Plant Domestication in Eastern North
        America. In *Rivers of Change: Essays on Early Agriculture in Eastern
        North America*, edited by Bruce D. Smith, pp. 19–34. Smithsonian Insti-
        tution Press, Washington, DC.

2001   Low-Level Food Production. *Journal of Archaeological Research* 9(1):
        1–43.

2007   Niche Construction and the Behavioral Context of Plant and Animal
        Domestication. *Evolutionary Anthropology* 16(5):188–199.

2011   The Cultural Context of Plant Domestication in Eastern North Amer-
        ica. *Current Anthropology* 52(S4):S471–S484.

2014   Failure of Optimal Foraging Theory to Appeal to Researchers Working
        on the Origins of Agriculture Worldwide. *Proceedings of the National
        Academy of Sciences* 111(28):E2829.

2015 A Comparison of Niche Construction Theory and Diet Breadth Models as Explanatory Frameworks for the Initial Domestication of Plants and Animals. *Journal of Archaeological Research* 23(3):215–262.

Smith, Bruce D., and C. Wesley Cowan

1987 Domesticated Chenopodium in Prehistoric Eastern North America: New Accelerator Dates from Eastern Kentucky. *American Antiquity* 52(2):355–357.

Smith, Bruce D., and Richard A. Yarnell

2009 Initial Formation of an Indigenous Crop Complex in Eastern North America at 3800 B.P. *Proceedings of the National Academy of Sciences* 106(16):6561–6566.

Smith, Eric A., Monique Borgerhoff Mulder, and Kim Hill

2001 Controversies in the Evolutionary Social Sciences: A Guide for the Perplexed. *Trends in Ecology and Evolution* 16(3):128–135.

Smith, Eric A., and Bruce Winterhalder

1992 Natural Selection and Decision-making: Some Fundamental Principles. In *Evolutionary Ecology and Human Behavior*, edited by Eric Alden Smith and Bruce Winterhalder, pp. 25–60. Aldine de Gruyter, New York.

Smith, Maria O.

1995 Scalping in the Archaic Period: Evidence from the Western Tennessee Valley. *Southeastern Archaeology* 14(1):60–68.

1996 Bioarchaeological Inquiry into Archaic Period Populations of the Southeast: Trauma and Occupational Stress. In *The Archaeology of the Mid-Holocene Southeast*, edited by Kenneth E. Sassaman and David G. Anderson, pp. 134–154. University Press of Florida, Gainesville.

Stark, Barbara

1986 Origins of Food Production in the New World. In *American Archaeology: Past and Future: A Celebration of the Society for American Archaeology, 1935–1985*, edited by David J. Meltzer, Don D. Fowler and Jeremy A. Sabloff, pp. 277–321. Smithsonian Institution Press, Washington, DC.

Steponaitis, Vincus P.

1986 Prehistoric Archaeology of the Southeastern U.S. *Annual Review of Anthropology* 15:383–393.

Stiner, Mary C.,

1990a Ungulate Exploitation during the Terminal Mousterian: The Case of Grotta Breuil. *Quaternaria Nova* 1:333–350.

1990b The Use of Mortality Patterns in Archaeological Studies of Hominid Predatory Adaptations. *Journal of Anthropological Archaeology* 9: 305–351.

1991a Actualistic and Archaeological Studies on Prey Mortality. In *Human Predators and Prey Mortality*, edited by Mary C. Stiner, pp. 1–13. Westview, Boulder, CO.

1991b An Interspecific Perspective on the Emergence of the Modern Human Predatory Niche. In *Human Predators and Prey Mortality*, edited by Mary C. Stiner, pp. 149–185. Westview, Boulder.

1993   Small Animal Exploitation and Its Relation to Hunting, Scavenging, and Gathering in the Italian Mousterian. In *Hunting and Animal Exploitation in the Later Palaeolithic and Mesolithic of Eurasia*, edited by Gail Larsen Peterkin, Harvey M. Bricker and Paul Mellars, pp. 101–119. Archaeological Papers No. 4, American Anthropological Association, Washington, DC.

1994   *Honor among Thieves: A Zooarchaeological Study of Neandertal Ecology.* Princeton University Press, Princeton.

2001   Thirty Years on the "Broad Spectrum Revolution" and Paleolithic Demography. *Proceedings of the National Academy of Sciences* 98(13): 6993–6996.

Stiner, Mary C., and Gillian Feeley-Harnik

2011   Energy and Ecosystems. In *Deep History: The Architecture of Past and Present*, edited by Andrew Shryock and Daniel Lord Smail, pp. 78–102. University of California Press, Berkeley.

Stiner, Mary C., Steven L. Kuhn, Stephen Weiner, and Ofer Bar-Yosef

1995   Differential Burning, Recrystallization, and Fragmentation of Archaeological Bone. *Journal of Archaeological Science* 22(2):223–237.

Stiner, Mary C., Steven L. Kuhn, Todd A. Surovell, Paul Goldberg, Liliane Meignen, Stephen Weiner, and Ofer Bar-Yosef

2001   Bone Preservation in Hayonim Cave (Israel): A Macroscopic and Mineralogical Study. *Journal of Archaeological Science* 28(6):643–659.

Stiner, Mary C., Natalie D. Munro, and Todd A. Surovell

2000   The Tortoise and the Hare: Small-Game Use, the Broad-Spectrum Revolution, and Paleolithic Demography. *Current Anthropology* 41(1): 39–79.

Stiner, Mary C., Natalie D. Munro, Todd A. Surovell, Eitan Tchernov, and Ofer Bar-Yosef

1999   Paleolithic Population Growth Pulses Evidenced by Small Animal Exploitation. *Science* 283(5399):190–194.

Straus, Lawrence, and Ted Goebel

2011   Humans and Younger Dryas: Dead End, Short Detour, or Open Road to the Holocene? *Quaternary International* 242(2):259–261.

Styles, Bonnie W., and Walter Klippel
1996   Mid-Holocene Faunal Exploitation in the Southeastern United States. In *Archaeology of the Mid-Holocene Southeast*, edited by Kenneth E. Sassaman and David G. Anderson, pp. 115–133. University Press of Florida, Gainesville.

Surovell, Todd A.
2009   *Toward a Behavioral Ecology of Lithic Technology: Cases from Paleoindian Archaeology*. University of Arizona Press, Tucson.

Surovell, Todd A., and P. Jeffrey Brantingham
2007   A Note on the Use of Temporal Frequency Distributions in Studies of Prehistoric Demography. *Journal of Archaeological Science* 34:1868–1877.

Surovell, Todd A., Judson Byrd Finely, Geoffrey M. Smith, P. Jeffrey Brantingham, and Robert L. Kelly
2009   Correcting Temporal Frequency Distributions for Taphonomic Bias. *Journal of Archaeological Science* 36:1715–1724.

Surovell, Todd A., and Nicole M. Waguespack
2008   How Many Elephant Kills Are 14? Clovis Mammoth and Mastodon Kills in Context. *Quaternary International* 191:82–97.

Sutherland, William J.
1996   *From Individual Behaviour to Population Ecology*. Oxford University Press, Oxford.

Thaler, Richard H.
2016   *Misbehaving: The Making of Behavioral Economics*. W. W. Norton, New York.

Thomas, Cyrus
1894   Report on the Mound Explorations of the Bureau of Ethnology. In *12th Annual Report of the Bureau of Ethnology to the Secretary of the Smithsonian Institution, 1890–'91*, pp. 3–742. Smithsonian Institution, Washington, DC.

Thomas, David Hurst
1975   Nonsite Sampling in Archaeology: Up the Creek Without a Site? In *Sampling in Archaeology*, edited by James W. Mueller, pp. 61–81. University of Arizona Press, Tucson.

1983   *The Archaeology of Monitor Valley: 2. Gatecliff Shelter*. Anthropological Papers Vol. 59, Pt. 1. American Museum of Natural History, New York.

Thomas, David Hurst, and Susan L. Bierwirth
1983   Material Culture of Gatecliff Shelter: Projectile Points. In *The Archaeology of Monitor Valley: 2. Gatecliff Shelter*, by David Hurst Thomas, pp. 177–211. Anthropological Papers Vol. 59, Pt. 1. American Museum of Natural History, New York.

Thomas, David Hurst, and Deborah Mayer
1983   Behavioral Faunal Analysis of Selected Horizons. In *The Archaeology of Monitor Valley: 2. Gatecliff Shelter*, by David Hurst Thomas, pp. 353–391. Anthropological Papers Vol. 59, Pt. 1. American Museum of Natural History, New York.

Thulman, David K.
2006   "A Reconstruction of Paleoindian Social Organization in North Central Florida." Ph.D. dissertation, Department of Anthropology, Florida State University, Tallahassee.
2012   Discriminating Paleoindian Point Types from Florida Using Landmark Geometric Morphometrics. *Journal of Archaeological Science* 39: 1599–1607.

Titmus, Gene L., and James C. Woods
1986   An Experimental Study of Projectile Point Fracture Patterns. *Journal of California and Great Basin Anthropology* 8(1):37–49.

Torrence, Robin
1983   Time Budgeting and Hunter-Gatherer Technology. In *Hunter-Gatherer Economy in Prehistory: A European Perspective*, edited by Geoff Bailey, pp. 11–12. Cambridge University Press, Cambridge.

Tremayne, Andrew H., and Bruce Winterhalder
2017   Large Mammal Biomass Predicts the Changing Distribution of Hunter-Gatherer Settlements in Mid–Late Holocene Alaska. *Journal of Anthropological Archaeology* 45:81–97.

Tukey, John W.
1953   *The Problem of Multiple Comparisons.* Unpublished manuscript, Princeton University, Princeton.

Tune, Jesse W.
2015   "Settling into the Younger Dryas: Human Behavioral Adaptations during the Pleistocene to Holocene Transition in the Midsouth United States." Ph.D. dissertation, Department of Anthropology, Texas A&M University, College Station.

Tylor, Edward
1871   *Primitive Culture.* J. P. Putnam's Sons, New York.

Ugan, Andrew, Jason Bright, and Alan Rogers
2003   When is Technology Worth the Trouble? *Journal of Archaeological Science* 30:1315–1330.

Ungar, Peter S., and Mark F. Teaford
2002   Perspectives on the Evolution of Human Diet. In *Human Diet: Its Origin and Evolution*, edited by Peter S. Ungar and Mark F. Teaford, pp. 1–6. Greenwood, Westport, CT.

U.S. Census Bureau

2010   *Population of the United States by County.* Electronic database, http://
       www.census.gov/2010census/, accessed May 2013.

U.S. Department of Agriculture, Natural Resources Conservation Service
(USDA-NRCS)

2013   *National Resources Conservation Service Official Soil Series Descriptions.*
       Electronic database, http://www.nrcs.usda.gov/wps/portal/nrcs/detail
       /national/soils/?cid=nrcs142p2_053627, accessed July 2013.

U.S. Geological Survey (USGS)

2013a *Generalized Geologic Map of the United States.* Electronic database,
       https://pubs.usgs.gov/atlas/geologic/, accessed March 2014.

2013b *North American Land Cover Characteristics—1 Kilometer Resolution.*
       Electronic database, https://nationalmap.gov/small_scale/mld/landcvi
       .html, accessed March 2013.

VanDerwarker, Amber M., and Tanya M. Peres

2010   Introduction. In *Integrating Zooarchaeology and Paleoethnobotany:
       A Consideration of Issues, Methods, and Cases,* edited by Amber M.
       VanDerwarker and Tanya M. Peres, pp. 1–15. Springer, New York.

Viau, A. E., K. Gajewski, M. C. Sawada, and P. Fines

2006   Millennial-Scale Temperature Variations in North America during the
       Holocene. *Journal of Geophysical Research* 111(D9):1–12.

Waguespack, Nicole M., Todd A. Surovell, Allen Denoyer, Alice Dallow, Adam
Savage, Jamie Hyneman, and Dan Tapster

2009   Making a Point: Wood- versus Stone-tipped Projectiles. *Antiquity* 83:
       786–800.

Walker, Renee B.

1998   "Late Paleoindian through Middle Archaic Faunal Evidence from Dust
       Cave, Alabama." Ph.D. dissertation, Department of Anthropology,
       University of Tennessee, Knoxville.

2007   Hunting in the Late Paleoindian Period: Faunal Remains from Dust
       Cave, Alabama. In *Foragers of the Terminal Pleistocene,* edited by
       Renee B. Walker and Boyce N. Driskell, pp. 99–115. University of
       Alabama Press, Tuscaloosa.

Walker, Renee B., Kandice R. Detwiler, Scott C. Meeks, and Boyce N. Driskell

2001   Berries, Bones, and Blades: Reconstructing Late Paleoindian Subsis-
       tence Economies at Dust Cave, Alabama. *Midcontinental Journal of
       Archaeology* 26(2):169–197.

Waters, Michael R., and Thomas W. Stafford, Jr.

2007   Redefining the Age of Clovis: Implications for the Peopling of the
       Americas. *Science* 315(5815):1122–1126.

Waters, Michael R., Thomas W. Stafford, Jr., Brian G. Redmond, and
Kenneth B. Tankersley
2009 The Age of the Paleoindian Assemblage at Sheriden Cave, Ohio. *American Antiquity* 74(1):107–111.

Webb, William S.
1939 *An Archaeological Survey of Wheeler Basin on the Tennessee River in Northern Alabama*. Bureau of American Ethnology Bulletin No. 122, Smithsonian Institution, Washington, DC.

Webb, William S., and David L. DeJarnette
1942 *An Archaeological Survey of Pickwick Basin in the Adjacent Portions of the States of Alabama, Mississippi, and Tennessee*. Bureau of American Ethnology Bulletin No. 129, Smithsonian Institution, Washington, DC.

Weitzel, Elic M., and Brian F. Codding
2016 Population Growth as a Driver of Initial Domestication in Eastern North America. *Royal Society Open Science* 3(8):160319.

Wilkins, Gary R., Paul A. Delcourt, Hazel R. Delcourt, Frederick W. Harrison, and Manson R. Turner
1991 Paleoecology of Central Kentucky since the Last Glacial Maximum. *Quaternary Research* 36(2):224–239.

Williams, Alan N.
2012 The Use of Summed Radiocarbon Probability Distributions in Archaeology: A Review of Methods. *Journal of Archaeological Science* 39(3):578–589.
2013 A New Population Curve for Prehistoric Australia. *Proceedings of the Royal Society B* 280:20130486.

Williams, John W., Bryan N. Shuman, Thomas Webb III, Patrick J. Bartlein, and Phillip L. Leduc
2004 Late-Quaternary Vegetation Dynamics in North America: Scaling from Taxa to Biomes. *Ecological Monographs* 74(2):309–334.

Wilson, Jennifer Keeling, and William Andrefsky, Jr.
2008 Exploring Retouch on Bifaces: Unpacking Production, Resharpening, and Hammer Type. In *Lithic Technology: Measures of Production, Use, and Curation*, edited by William Andrefsky, Jr., pp. 86–105. Cambridge University Press, Cambridge.

Winterhalder, Bruce, and Carol Goland
1997 An Evolutionary Ecology Perspective on Diet Choice, Risk, and Plant Domestication. In *People, Plants, and Landscapes: Studies in Paleoethnobotany*, edited by Kristen J. Gremillion, pp. 123–160. University of Alabama Press, Tuscaloosa.

Winterhalder, Bruce, Douglas J. Kennett, Mark N. Grote, and Jacob Bartruff
2010   Ideal Free Settlement of California's Northern Channel Islands. *Journal of Anthropological Archaeology* 29:469–490.

Winterhalder, Bruce, Cedric Puleston, and Cody Ross
2015   Production Risk, Inter-Annual Food Storage by Households and Population-Level Consequences in Seasonal Prehistoric Agrarian Societies. *Environmental Archaeology* 20(4):337–348.

Winterhalder, Bruce, and Eric Alden Smith
2000   Analyzing Adaptive Strategies: Human Behavioral Ecology at Twenty-Five. *Evolutionary Anthropology* 9(2):51–72.

Wobst, H. Martin
1978   The Archaeo-Ethnology of Hunter-Gatherers or the Tyranny of the Ethnographic Record in Archaeology. *American Antiquity* 43(2): 303–309.

Woodburn, James
1982   Egalitarian Societies. *Man* 17(3):431–451.

Yarnell, Richard A.
1978   Domestication of Sunflower and Sumpweed in Eastern North America. In *The Nature and Status of Ethnobotany*, edited by Richard I. Ford, pp. 289–299. Anthropological Papers No. 67, Museum of Anthropology, University of Michigan, Ann Arbor.

Zeanah, David W.
2016   Foraging Models, Niche Construction, and the Eastern Agricultural Complex. *American Antiquity* 82(1):3–24.

Zeder, Melinda A.
2011   The Origins of Agriculture in the Near East. *Current Anthropology* 52(S4):S221–S235.

2012   The Broad Spectrum Revolution at 40: Resource Diversity, Intensification, and an Alternative to Optimal Foraging Explanations. *Journal of Anthropological Archaeology* 31:241–264.

2015   Core Questions in Domestication Research. *Proceedings of the National Academy of Sciences* 112(11):3191–3198.

Index

Note: Page numbers in *italics* refer to figures or tables.